Technical Milestones

Englisch für technische Berufe

Workbook mit Audio-CD-ROM

von
Wolfram Büchel
Cornelius Carey
Mary Schäfer
Dr. Wolfgang Schäfer

Ernst Klett Verlag
Stuttgart · Leipzig

Technical Milestones
Englisch für technische Berufe
Workbook mit Audio-CD-ROM

Autorinnen und Autoren: Wolfram Büchel, Cornelius Carey, Mary Schäfer, Dr. Wolfgang Schäfer

Werkübersicht:

Schülerbuch, 978-3-12-808268-4
Workbook mit Audio-CD-ROM, 978-3-12-808269-1
Lehrerhandbuch inkl. Digitalem Lehrer-Service, 978-3-12-808270-7

1. Auflage

1 $^{7\ 6\ 5\ 4\ 3}$ | 26 25 24 23 22

Alle Drucke dieser Auflage sind unverändert und können im Unterricht nebeneinander verwendet werden.
Die letzte Zahl bezeichnet das Jahr des Druckes.

Im Lehrwerk befinden sich ausschließlich fiktive Internet-Adressen, die deshalb auch mit ww#. beginnen anstatt wie üblich mit www.

Projektleitung: Karin Altrichter, Matthias Rupp
Redaktion: Vera Kirchhübel
Herstellung: Sarah Ganser, Angelika Lindner

Satz: Beckers Büro, Claudia Becker, Stuttgart
Umschlaggestaltung: Projektteam des Verlages
Reproduktion: Meyle + Müller Medien-Management, Pforzheim
Druck: Druckhaus Götz GmbH, Ludwigsburg

Tontechnik: Klett Studio, Stuttgart und Bauer Studios, Ludwigsburg
Presswerk: Osswald GmbH & Co., Leinfelden-Echterdingen

Printed in Germany
ISBN 978-3-12-808269-1

Verweise

⊚ **A 1.17**   CD 1, Track 17 im Lehrerhandbuch; Track 1_17 auf der Workbook CD-ROM

→   Grammatikanhang des Schülerbuches

Das Workbook ermöglicht es Ihnen, die im Schülerbuch erlernten und geübten Themenbereiche und Kompetenzen zu vertiefen und zu festigen. Die Module des Workbooks orientieren sich an denen des Schülerbuchs. Das Workbook kann allerdings auch ohne den Schülerband verwendet werden.

Jedes Modul ist in folgende Bereiche unterteilt:

1. Module refresher

In diesem Bereich werden die Themenfelder des Schülerbuches in Hörverständnis-übungen zu authentischen Dialogen, welche auf der beiliegenden Audio-CD-ROM zu finden sind, sowie Vokabel- und Mediationsübungen vertiefend geübt.

2. Grammar refresher

In diesem Teil werden die im Schülerbuch vermittelten grammatikalischen Strukturen zusammenfassend dargestellt. In verschiedenen Aufgabentypen werden diese wiederholt und gefestigt.

3. Phrases refresher

Ein grundlegendes Repertoire an Redemitteln ist für erfolgreiches Kommunizieren in technischen Berufen essenziell. In diesem Sinne bietet das Workbook zu den *Phrases* aus dem Schülerbuch weitere Übungsmöglichkeiten an.

Im Anschluss an die Module können berufsbezogene Fremdsprachenkenntnisse in Aufgaben, die auf das KMK-Fremdsprachenzertifikat vorbereiten, intensiviert werden. Die Bedeutung dieser Zertifizierung wird von vielen Arbeitgebern betont, weil die immer tiefgreifendere Globalisierung Fremdsprachen in der Arbeitswelt unverzichtbar werden lässt. Nähere Informationen zum KMK-Fremdsprachenzertifikat geben die Hinweise vor der KMK-Prüfungsvorbereitung.

A Module refresher

1 Complete the following sentences. You will find the missing words in the word search below.

A	D	B	N	M	I	T	I	S	A	T	I	I	S	A	R	T
B	U	C	H	A	R	E	S	T	R	H	A	D	O	Q	N	O
E	T	R	I	D	O	V	W	H	E	P	F	I	M	L	D	A
R	C	L	K	R	O	M	E	O	N	O	R	W	A	Y	O	B
I	H	V	T	I	S	A	D	E	K	L	K	A	P	S	H	A
A	G	Y	R	D	P	F	I	N	N	I	S	H	I	N	I	T
R	H	L	K	F	O	W	S	G	H	S	O	U	J	E	K	O
I	O	N	U	E	D	W	H	I	L	H	I	N	T	A	P	F
F	W	J	A	V	B	G	E	Q	M	O	H	G	O	B	H	A
A	W	V	R	K	R	B	A	P	N	K	G	A	H	O	K	F
D	H	G	Y	L	A	N	Q	V	H	E	P	R	A	G	U	E
I	O	F	I	D	E	U	K	I	J	M	L	I	V	M	V	G
Y	O	M	A	D	R	U	B	R	E	A	O	A	S	U	P	Q
A	R	T	O	P	K	M	A	T	I	N	L	N	K	O	M	A

1. … is the capital of Romania.
2. Sophie is from Sweden. She is … .
3. Raija lives in Helsinki. She speaks … .
4. People who live in the Netherlands speak … .
5. The capital of Spain is … .
6. Most people who live in Warsaw are … .
7. The main language spoken in Budapest is … .
8. The capital of Italy, … , is known as the eternal city.
9. Oslo is the capital of … , which is not a member of the EU.
10. Václav lives in the capital of the Czech Republic, … .

2 You have just arrived at the Romanian branch of your company and you are telling your Romanian colleague how your trip was. He speaks English but does not understand every word you say. Explain the terms below to him in one or two English sentences. The words in the box may help you.

> to serve • passage • to sell • exhibition • suitcase • to present • passengers • rows of seats

1. aisle: _____

2. sample case: _____

3. trade fair: _____

4. flight attendant: _____

3 Listen to the dialogue and complete the conversation.

A 1.2

Helen: There isn't any time for a **1** _____ today; the flight is too short. And I'm so **2** _____ .

Oliver: Me, too … . I don't know much about English **3** _____ . What can you recommend?

Helen: Well, you can't go wrong with **4** _____ , and the English love their **5** _____ .

Oliver: I expect to be in Leeds for at least a year. You said you're from Leeds. Do you know any good **6** _____ there?

Helen: There are lots of good places. Where's your company in Leeds?

Oliver: It's Power Engines in Russell Street.

Helen: That's only three streets from my office. We can meet for **7** _____ some day next week if you like. Here's my card.

Oliver: That's a good idea. Thanks very much.

4 Complete the sentences. Finish them any way you like; as long as they make sense.

1. Oliver is travelling to England _____

2. Helen doesn't like mobile phones in restaurants _____

3. Fernando thinks that in emergencies mobile phones _____

4. Helen likes her notebook _____

5. During takeoff or landing you are not allowed to _____

B Grammar refresher

> **Prepositions**
>
> Präpositionen *zeigen, wie Menschen und Sachen in Zeit und Raum zueinander stehen.*
> **Example:** Please give me the file that is lying **next to** the telephone.
> → **SB p. 234 Prepositions**

1 Complete the safety instructions given in an aircraft using suitable prepositions.

"Please place all hand luggage **1** _____ your seat. Anything you do not need during the flight

should be put **2** _____ the compartment **3** _____ you. Please do not stand

4 _____ the aisle or **5** _____ the toilets, unless you are waiting to use them. You will find

the emergency exits **6** _____ row 13. If the emergency sign goes **7** _____ , listen

8 _____ the captain's announcement. An oxygen mask will drop **9** _____ the panel

10 _____ your head. There is a life jacket **11** _____ each seat. Take it **12** _____

and put it **13** _____ and then pull **14** _____ the cord to inflate the jacket."

> **to be and personal pronouns**
>
> | singular | plural | Example: **She is** from Spain. |
> |---|---|---|
> | I am | we are | |
> | you are | you are | |
> | he / she / it is | they are | |
>
> → **SB pp. 9, 11 Forms of to be – present tense**

2 Complete the sentences by filling in the missing forms of to be and / or personal pronouns.

There **1** _____ five people in the first row of the plane. All of the friends **2** _____ there except

Fernando. **3** _____ sitting all by himself in the second row. "Unfortunately, **4** _____ weren't able

to fit **5** _____ in the first row together," apologises the stewardess. "Don't worry," says Oliver,

"**6** _____ will **7** _____ fine."

> **Much, many, little, few**
>
> Much, a lot of, a great deal of *und* little *werden für nicht zählbare Nomen verwendet.* Many
> *und* few *werden für zählbare Nomen im Plural verwendet.*
> **Example: Many** planes are delayed due to the weather; very **few** are able to take off.
> → **SB p. 13 Much, many, little, few**

3 Complete the sentences using much, many, little or few.

1. This is a business trip; I have _____ interest in sightseeing.

2. You won't have _____ problems with English. No one will mind if you make a _____ mistakes.

3. I don't know _____ about Ireland, but _____ of my colleagues have been there.

> **Some and any**
>
> Some *wird in bejahten und* any *in verneinten Sätzen, Fragen, oder wenn jemand Zweifel ausdrückt, verwendet.*
> **Example:** Here are **some** newspapers; we haven't **any** with the latest election results.
> → **SB p. 13 Some and any**

4 **Complete the sentences using some and any.**

1. I hope we will have _____ time to get a coffee before takeoff.

2. Do you have _____ thing for me to read? – We might even have _____ German newspapers.

3. Do you serve _____ French wine? – I'm afraid we don't have _____ French wine today.

> **Past Tense**
>
> *Das* Past Tense *wird für Geschehnisse oder Aktivitäten, die abgeschlossen sind, verwendet.*
> **Example:** I **lived** in England for three months in 2010.
> → **SB p. 224 Past Tense**

5 **Put the sentences below into English using the past tense.**

1. Oliver war in der Schule gut in Mathematik.

   _____

2. Er arbeitete bei *(work with)* der Firma Hieke CNC Technik in Hannover.

   _____

3. Er wollte seine Englischkenntnisse verbessern.

   _____

4. Er entschloss sich für ein Jahr in England zu arbeiten.

   _____

C Phrases refresher

Complete the conversation using suitable phrases.

Oliver Klein	Rodrigo Álvarez
1 _____ , would you be so kind as to get my coat from the luggage compartment?	**2** _____ , here you are. By the way, **3** _____ Rodrigo Álvarez. **4** _____ Malaga. I am visiting my brother in Manchester.
5 _____ Oliver Klein. I am going to work in England for a year.	**6** _____ , Mr Klein.
7 _____ Oliver.	And I am Rodrigo.

Module 2 The new company

A Module refresher

1 First read the text below. Then listen to the CD and complete the conversation.

A 1.5

First Oliver and Bill greet each other, then Bill asks:

Bill: Did you have a **1** _____ flight?

Oliver: Yes, thank you, **2** _____ there was a bit of a **3** _____ .

After having offered something to drink, Bill continues:

Bill: Right, let me introduce you to your new **4** _____ . This is Roberto Garcia

from **5** _____ .

Oliver: **6** _____ to meet you.

Roberto: **7** _____ to meet you, **8** _____ .

Finally Oliver meets Steven Hill:

Bill: Ah, there's Steven. Oliver, **9** _____ I **10** _____ you to

your new boss? This is Steven Hill, who is **11** _____ of production,

12 _____ , and **13** _____ .

Steven: How do you do? Pleased to meet you and

14 _____ to Power

15 _____ . I hope you will

16 _____ working for us.

Oliver: I'm **17** _____ I will.

Bill: Before Steven shows you around the

18 _____ workshops,

I'd like to give you a **19** _____

outline of our company …

2 Hören Sie sich eine Führung durch die Firma Power Engines Ltd. an. Ein Besucher dieser Gruppe versteht kein Englisch. Fassen Sie kurz auf Deutsch für ihn zusammen, wer die Mitarbeiterinnen und Mitarbeiter von Power Engines sind, die die Gruppe auf der Tour begleiten und was ihre Aufgabe bei dieser Tour ist.

A 1.7

Emily Miller _____

John Smiley _____

Peter Thompson _____

Steven Hill _____

Jane Owtrim _____

Silke Meier _____

B   Grammar refresher

> **Comparison of adjectives**
>
> *Bei* einsilbigen Adjektiven *wird -er und -est gesteigert (germanische Steigerung).* Zweisilbige, *die auf -y, -le, -er oder -ow enden werden ebenso germanisch gesteigert.* Andere zweisilbige *und* alle drei- und mehrsilbigen Adjektive *werden* romanisch *gesteigert.*
> **Examples:** bright, brighter, the brightest; clever, cleverer, the cleverest;
> adventurous, more adventurous, the most adventurous
> *Es gibt auch* unregelmäßige Adjektive:
> good, better, the best; bad, worse, the worst; much / many, more, the most;
> little, less, the least; old, elder, the eldest (bei Familienangehörigen)
> → **SB p. 226 Comparison of adjectives**

1   **Put in the adjectives in brackets in the correct form.**

1. The new CNC centre is _____ (fast) and _____ (accurate) than the old one.

2. However, the new machine also seems _____ (complicated).

3. In their adverts the company calls it the _____ (good) CNC centre in the world.

4. Mr Baker is very _____ (pleased) with the new machine.

5. But he needs a special training course in order to get the _____ (much) out of the machine.

> **Adverbs**
>
> *Das* Adverb *dient zur näheren Beschreibung eines Verbs oder eines Adjektivs.*
> *Ein* Adverb *wird durch Anhängen von -ly an das* Adjektiv *gebildet:* quick – quick**ly**.
> *Wenn das* Adjektiv *selbst auf -y endet, wird dieses zu i:* happy – happily.
> *Ein stummes* e *entfällt:* true – truly; *ein Endkonsonant wird verdoppelt:* careful – carefully.
> Adverbien *können auch gesteigert werden:* carefully – more carefully – most carefully.
> **Examples:** The machine works **well**.
> The small printer makes **remarkably** clear printouts.
> Our new machine works **extremely** quickly.
> → **SB p. 227 Adverbs**

2   **Complete the presentation at a company meeting using adjectives and adverbs. You may use a dictionary.**

"Power Engines has developed **1** _____ this year (wonderful). This was a surprise because

we were confronted with a very **2** _____ market last year (competitive). Our sales had

fallen **3** _____ (sharp). Due to our new engine control system, sales began to

increase **4** _____ during the first months of this year (considerable). This produced a

5 _____ turnover for the first quarter of the year than expected (good). This improvement

is due to the new control system, which sells **6** _____ (exceptional / good).

Of course, this success was also **7** _____ facilitated by Ms McKenzie from marketing

(great). So the value of our shares on the stock market has risen **8** _____ (sharp).

However, costs must be watched **9** _____ to keep our prices **10** _____

(careful / competitive)."

C Phrases refresher

1 Read the situations carefully. Then tick the correct answers.

1. If you want to introduce yourself to another person, you can say (more than one is possible):
   - ☐ "My name is …"
   - ☐ "I'm called …"
   - ☐ "May I introduce myself, …"
   - ☐ "Mayer, Helmut"

2. Your best answer after somebody says "Nice to meet you" is (only one is possible):
   - ☐ "Thank you, that's nice."
   - ☐ "Nice to meet you, too."
   - ☐ "I'm happy to be here."

3. If you want to introduce a colleague to a visitor, which of the phrases are casual ways of introducing, which ones are very formal (more than one possibility)?
   - ☐ "Have you met Ms Saunders before?"
   - ☐ "Mr Miller, I would like you to meet our Ms Saunders."
   - ☐ "This is Ms Saunders."
   - ☐ "May I introduce Ms Saunders?"

4. Which phrase is <u>not</u> appropriate when you make a phone call and want to give your name?
   - ☐ "This is …"
   - ☐ "My name is …"
   - ☐ "John Smiley, I want to speak to …"

5. You didn't understand a caller's name. What do you say (more than one possibility)?
   - ☐ "What was your name again?"
   - ☐ "I don't understand, who are you?"
   - ☐ "Who did you say was calling?"

6. You are taking a call and the caller wants to speak to someone who is not in. You say (more than one possibility):
   - ☐ "I'll have him return your call."
   - ☐ "Would you like to leave a message?"
   - ☐ "Can I take a message?"

2 Sie arbeiten in der Einkaufsabteilung Ihres Betriebes und schreiben eine E-Mail an Max Silverstone. Passen Sie Datum und Termine an Ihr reales Datum an. Ein zweisprachiges Lexikon ist erlaubt. Folgende Punkte sollte Ihre E-Mail enthalten:

- Betreff und Anrede in der E-Mail wählen Sie selbst.
- Danken Sie ihm für die gestrige Mail.
- Danken Sie ihm auch für die im Anhang mitgeschickte Präsentation und die Zeichnungen, die Sie problemlos öffnen konnten.
- Sie möchten den Termin für das Treffen am nächsten Freitag bestätigen.
- Frau Metzler, ihre Produktionsleiterin, wird an dem Treffen teilnehmen.
- Schreiben Sie, dass Sie die Dokumentation vorbereitet haben und auch eine Kopie des neuen Vertrages mitbringen werden.
- Schließen Sie die Mail freundlich: Wünschen Sie z.B. ein schönes Wochenende/Sie freuen sich auf das Treffen nächste Woche.

3 Sie haben auf der CeBit in Hannover einen Prospekt der Firma Graphic Engines Ltd. mit dem neuartigen Flachbildschirm FlatCad X14 erhalten. Leider war am Stand der Firma niemand zu sprechen. Schreiben Sie deshalb für Ihre Firma SoftDesign einen Brief. Übertragen Sie hierfür die untenstehenden Informationen sinngemäß ins Englische.

SoftDesign

Carl-Tschamber-Straße 1
50667 Köln

Graphic Engines Ltd.
56 Colgrove Street
Loughborough, Leicestershire LE11
UK

Date

- Vor allem wollen Sie wissen, ob der Bildschirm für technische Zeichenprogramme geeignet ist, da Sie in Ihrer Firma Maschinenteile in 3D konstruieren und es nötig ist, dass Sie die Teile aus unterschiedlichen Winkeln betrachten können.
- Sie benötigen für Ihre Firma 15 Bildschirme und hoffen auf einen guten Preis. Erkundigen Sie sich auch nach den Liefer- und Zahlungsbedingungen.
- Schreiben Sie, dass die Informationen gerne über Internet geschickt werden können. Vergessen Sie nicht, entweder hier oder im Absender eine E-Mail Adresse anzugeben. Verbleiben Sie in der Hoffnung auf schnelle Beantwortung (Bleiben Sie hierbei höflich.). Unterzeichnen Sie als Leiter der Einkaufsabteilung.

4 Put in the missing words or phrases in the personal profile below.

My name is _____ .

I work for / I work with _____ (name of your company) in _____

(name of the city, town or village your company is based).

My company produces / works / deals with _____ .

The company has about _____ employees (number of people who are employed). I have worked for

my company for _____ years now. I work in the _____ department. What I enjoy

most here is _____ (e. g. the good atmosphere / interesting work).

I need English on the job because _____

(e.g. have to deal with customers / have to visit customers / am abroad a lot / am on trade fairs).

My family

I live in _____(name of city, town or village). _____ is a village /

town / city near _____ (name of bigger town or place).

I live _____

(by myself / with my wife / husband and ... children, a daughter named ... and a boy named ...).

My hobbies are _____ .

Module 3 Information technology

A Module refresher

1 Complete the sentences with suitable words.

1. Hardware is a term that covers all the _____ components of a computer.

2. Modern notebooks have excellent touch pads, but many computer users still prefer a _____ .

3. Most notebooks come with an on-board _____ card.

4. Hubs have been replaced by the much faster _____ .

5. The USB is probably the most common _____ device.

6. The _____ key stops a program or a command without losing data.

7. With the _____ key you can switch to a capital letter.

8. With a combination of the *Alt Gr key* and the key for the letter Q you can produce the _____ sign needed for email addresses.

9. A _____ is not just for computer games; you can control machines with it, and in the future you will probably even be able to steer a car with it.

2 Put the following sentences into the correct order.

1. The space bar / blank / it / you / produces / space / a / when / press

   _____

2. the page down / can / With / leaf through / page up / or / key/ a manuscript / page wise / you

   _____

3. single inventor / was no / the computer / of / was / There

   _____

4. Charles Babbage / In the 1830s / invented / calculator / mechanical / a

   _____

5. computers / You / but / should / shouldn't / control / you / they / control

   _____

3   **Put the sentences below into German.**

1. Charles Babbage invented a mechanical calculator in the 1830s. But Blaise Pascal and Wilhelm Leibniz also belong to the inventors of mechanical calculation machines.

   _____

   _____

2. German computer specialists who know something about the history of computers will tell you that Konrad Zuse built the first programmable computer in Berlin between 1936 and 1941.

   _____

   _____

3. In 1971 the first microprocessor appeared on the market.

   _____

4. When vacuum tubes were replaced by transistors, the development of computers got a new impetus.

   _____

   _____

5. Computers control too much of our daily life – that is what critics say.

   _____

6. One still doesn't know today which direction the internet will take.

   _____

7. Problem number 1 will undoubtedly be security.

   _____

8. One thing is definitely clear: the internet will take on more and more tasks.

   _____

9. The main advantage of model A over models B and C is that all of the hardware components in model A are really state-of-the-art whereas the other two models contain only the standard equipment.

   _____

   _____

   _____

   _____

   _____

   _____

B Grammar refresher

Modal auxiliaries

Modale Hilfsverben *sind z. B.* can, must, shall, should, may, might, need / needn't.
Vorsicht: *mustn't = nicht dürfen; needn't = nicht müssen.*
Example: You **needn't** do it. (= Du brauchst das nicht zu tun.)
Man nennt diese Verben auch unvollständige Hilfsverben (defective auxiliaries)*, weil sie nicht in alle Zeiten gesetzt werden können. Es gibt bei ihnen kein -s in der dritten Person:* he can, he should.
Zur Bildung der anderen Zeiten muss auf die Ersatzformen ausgewichen werden:
can – to be able to = *können, fähig sein, in der Lage sein*
can – to be allowed to = *dürfen, die Erlaubnis haben*
may – to be allowed to = *dürfen, die Erlaubnis haben*
must – to have to = *müssen*
mustn't – not to be allowed to = *nicht dürfen*
needn't – not to have to = *nicht brauchen, nicht müssen*
Example: He **had to** install the program again. (= Er musste das Programm erneut installieren.)
Did he **have to** reboot the system? (= Musste er das System neu starten?)
Vorsicht : He **could** start the program. (= Er **könnte** das Programm starten.)
He **couldn't** start the program. (= Er konnte das Programm nicht starten. ODER Er könnte das Programm nicht starten.)
→ **SB p. 235 Modal auxiliaries**

1 **Put the words in the box into the gaps below. Some of them can be used more than once.**

can • can't • should • shouldn't • must • mustn't • need to • needn't • might

You **1** _____ leave children unattended in front of the computer. As a parent you **2** _____

be in the same room with them, but you **3** _____ at least be around so your children

4 _____ ask for help if they need it.

You absolutely **5** _____ give away passwords or pin codes for your bank accounts.

You **6** _____ always be careful with cheap offers over the internet.

You **7** _____ decide for yourself whether you want to do bank transactions over the internet.

With emails you **8** _____ be careful, too. You **9** _____ be careful with attachments,

because they **10** _____ contain viruses or other malware. You **11** _____ open an email

that seems suspicious. And when you write emails yourself, think twice before you send them because

you **12** _____ take them back.

2 **Complete the sentences with the appropriate auxiliaries in the past tense.**

Example: A: Can you start your computer?

    A: **Were you able**   to start your computer last night?

A: Can you access the company website?

A: _____ **1** access the company website yesterday?

B: No, not at all, the message tells me that I may not access the site as I'm not the admin.

B: No, not at all, the message told me that

I _____ **2** access the site as I'm not the admin.

A: So, what do you have to do now?

A: So, what _____ **3** next?

B: I must call the admin.

B: Then I _____ **4** call the admin.

A: Any help from him?

B: He says I must reinstall the programme.

B: He said I _____ **5** reinstall the programme.

A: What a waste of time!

C   Phrases refresher

A customer of yours is looking for a new notebook. Present her some of the latest models using the phrases below. Make five sentences. There is often more than one possibility.

The Netway Bundle	**has a built-in**	a 5-Channel Sound card.
	comes with	WLAN card.
The Xentum Notebook	**is equipped with**	a 4 GHZ dual core CPU.
	features	the Cartellsoft Asta Professional Operating system.
The Infoflex Notebook	**has an advantage over**	the Infoflex Notebook, because it has a battery capacity of 4 hours.

_____

_____

_____

_____

_____

_____

_____

_____

_____

_____

_____

A Module refresher

1 Look at the lathe, milling machine and grinding machine below and label the machine parts with the corresponding machine tools in the pictures.

centre lathe

1 tailstock
2 headstock with gearbox
3 machine bed with slide
4 emergency stop
5 saddle and cross slide
6 tool quick exchange
7 handwheel control
8 toolpost
9 main spindle with chuck
10 control rod

milling machine

1 overarm
2 column
3 vertical slide
4 knee
5 control panel
6 machine table
7 base

grinding machine

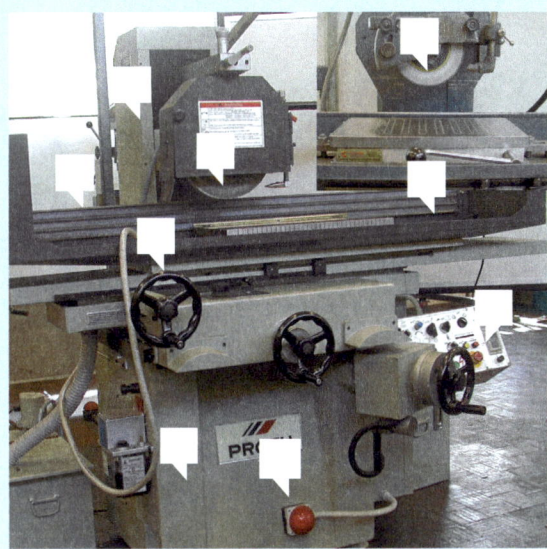

1 table traverse
2 column
3 table hand feed
4 grinding disc
5 base
6 machine table
7 emergency stop
8 start and stop switch

2 Complete the crossword by filling in the tools you need for the jobs below.

Down

1. Operating flat material with different forms
2. Bending a wire
4. Finishing the surface of flat metal after milling
5. Clamping workpieces for filing or deburring
8. Assembling sheet steel as a fixed joint

Across

3. Turning screws into wood or tin
6. Drilling or boring holes up to 12 mm
7. Driving nails into a wooden board
9. Tightening and loosening bolts and nuts
10. Tightening or loosening hexagon socket head bolts

3 Match the tools mentioned in the dialogue with the functions below.

A 2.12

1. It houses the gears of a lathe. _____

2. It feeds the cutting tool during the turning operation. _____

3. It holds the workpiece during the milling process. _____

4. It feeds the workpiece to the grinding wheel. _____

5. It holds the workpiece during the lathe cutting process. _____

6. It holds the lathe cutting tool. _____

B Grammar refresher

> **Questions with interrogatives**
>
> Fragen mit Fragefürwörtern (questions with interrogatives) *werden mit dem Fragefürwort eingeleitet. Vor dem Subjekt steht die Form von* to do, *dahinter steht das* Vollverb im Infinitiv.
> Example: Liz **checked** the **milling machine. What did** Liz **check?**
> *Wird nach einem Teil des Subjekts gefragt (who, what, which, whose), so wird die Form von* to do *nicht verwendet.*
> Example: **Liz checked** the milling machine. **Who checked** the milling machine?
> → **SB p. 228 Questions**

1 Ask questions about the words in blue.

Example: Helen saw her instructor in the CNC-shop of the training centre.

Who did Helen see in the CNC-shop of the training centre?

1. Mr Miller showed his visitors the shop floor of the company.

   _____

2. The trainees get training on the machine tools because they'll be operating them later.

   _____

3. You can see a column drill between the windows.

   _____

4. The milling machine will be set up next month.

   _____

5. The riveting tongs are used to join sheet metals.

   _____

6. The main spindle rotates the three-jaw chuck with the workpiece.

   _____

7. The machine starts running after the worker has pushed the start and stop button.

   _____

> **Relative clauses**
>
> **Defining relative clauses:**
> *Ein* Relativsatz *ist notwendig* (defining relative clause), *wenn die darin enthaltene Information nötig ist, um deutlich zu machen, welche Person oder Sache gemeint ist. Bei Personen wird* who *verwendet, bei Sachen* which *oder* that. *Steht das* Relativpronomen im Akkusativ, *so kann es weggelassen werden.*
> Examples: The young man **who** is cleaning the lathe works in the milling shop.
> Ben studies the technical drawing **(which)** he designed on the computer before.
> **Non-defining relative clauses:**
> *Ein* Relativsatz *ist nicht notwendig* (non-defining relative clause), *wenn die Information nicht nötig ist, um zu zeigen, welche Person / Sache gemeint ist.*
> Example: The milling machine, **which** is used for training, always runs perfectly.
> → **SB p. 230 Defining relative clauses**
> → **SB p. 230 Non-defining relative clauses**

2 Use the correct relative clauses to connect the following sentences.

1. The visitors to the company are asked not to take photos. The visitors are from abroad.

   _____

2. The instructor is preparing a training sequence for his apprentices. He is presently setting up a lathe.

   _____
   _____

3. The machine vice is mounted on the machine table. It holds the material in the position required.

   _____

4. Alexander needs some help with his milling job. The milling job is described in the work plan.

   _____

5. The turner must exchange the reversible carbide tip for the cutting operation. The turner is highly skilled for machining processes.

   _____
   _____

6. The latest lathe model is equipped with a high-speed spindle. It was demonstrated to visitors.

   _____

7. The finishing process of the milled part is prepared by Sven. He is in his last year of training.

   _____

8. The firm's visitors liked the demonstration. The demonstration was about CNC machining processes.

   _____

C Phrases refresher

1 Complete the following sentences of a product advertisement.

1. This is our _____ model of a CNC milling machine.

2. It has the interesting new _____ of a high speed turret automatic (=Revolverautomat).

3. The major _____ is that regular machine service is included.

4. Another _____ of this model is the tool capacity of the turret.

5. This means that the new model _____ better than the previous one.

6. The control panel is very _____ for programming all required machining processes.

2 Write down the safety instructions for the following situations.

1. The trainees are going to weld some metal bars.

   _____

2. The worker has to carry heavy stock from the storage room into the workshop.

   _____

3. The electrician wants to repair the cable of a power drill.

   _____

4. The apprentices are going to grind the surface of a metal block.

   _____

Module 5 Joining and assembly

A Module refresher

1 Match the different joining methods below with the corresponding category in the boxes.

welding • riveting • bonding • screwing • fitting • lasing • soldering • nailing

Mechanical joint	Non-mechanical joint
–	–
–	–
–	–
–	–

2 Look at the illustrations and list the components used for either welding or soldering.

gas welding gear

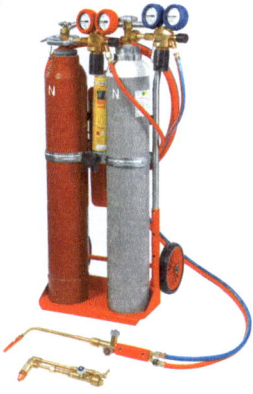

soldering station

_____

_____

_____

3 Write down the tools which are used for the various joining methods listed below.

welding	–
riveting	–
bonding	–
screwing	–
fitting	–
lasing	–
soldering	–
nailing	–

4 Two trainees are discussing various joining techniques and the corresponding tools applied for the jobs below. Complete the dialogue as shown in the example using the following phrases:

Do you remember how to … ?   Do you know how … ?   Which tool is used to … ?   How do you … ?

Example: join two pieces of sheet metal

A: **Which tool is used to join two pieces of sheet metal?**

B: **Well, you use riveting tongs and rivets to join two pieces of sheet metal.**

1. assemble two base plates of a machine vice

   _____

2. repair the broken handle of a tea cup

   _____

3. join flat steel plates to form a square

   _____

4. mount the aluminium shelves in the store room

   _____

5. fix the copper wire to the circuit board

   _____

6. hang a picture on the wall

   _____

7. assemble a wooden book shelf

   _____

5 Listen to the dialogue and list the parts of the welding equipment mentioned.

A 1.17

1. _____   2. _____

3. _____   4. _____

5. _____

6 Match the English and the German words for protective clothing.

1. feuerfeste Stulpenhandschuhe	a) welding hood with filtered lens
2. feuerfeste Schweißerjacke	b) fire-resistant jacket
3. Lederschürze	c) fire-resistant gauntlet (=Stulpen) welding gloves
4. Schweißerschutzmaske mit Filterbrillenglas	d) leather apron

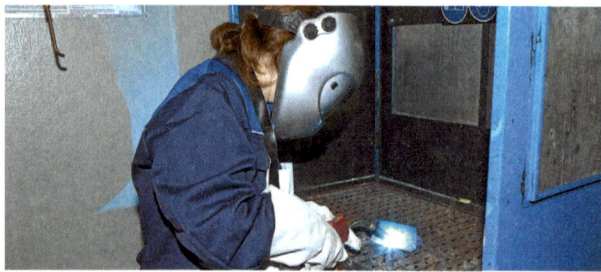 

B Grammar refresher

Passive voice

Das Passiv *wird gerne im technischen Englisch benutzt, wenn technische Vorgänge beschrieben werden.*
Das Passiv *wird im Englischen gebildet mit einer Form von* to be *und der dritten Stammform des Verbes (Partizip Perfekt).*
Examples: Before the machining process **is started**, the mains **are switched** on.
A blunt drill **cannot be used** if it **has not been sharpened.**
→ **SB p. 232 Passive voice**

1 Complete the following sentences by filling in the correct passive form.

1. For each joining method different tools _____ (require).

2. Spanners, for example, _____ (use) to tighten or loosen bolts and nuts.

3. A welding torch as well as gas and oxygen cylinders _____ (need) for welding jobs.

4. Cross-slot screws should _____ (drive in) with cross-slot screwdrivers.

5. When rivets are used for joining processes, the rivets _____ (shoot) out of a gun.

6. The bolts _____(fasten) with nuts after they have passed through the metal.

7. To produce a safer joint, a washer _____ (insert) between the bolt and the nut.

8. In welding the oxygen _____ (mix) with the acetylene to increase the temperature of the welding flame.

9. The flame can _____(adjust) to reach the right temperature.

10. The pressure in the gas and oxygen cylinders _____ (regulate) with the regulator gauges.

2 Describe how a locking seam (=Falzverbindung) is folded. Rewrite the following instructions of a manual using the passive voice.

How to fold a locking seam

1. You can join sheet metal in different ways.

   _____

2. The worker scribes lines on the sheet metal where he wants the folding edges.

   _____

3. He clamps the first sheet in the vice.

   _____

4. He taps the edges to a 90-degree angle.

   _____

5. Then he bends the edge to almost 180 degrees.

   _____

6. He repeats the process with the second sheet of metal.

   _____

7. Finally, he hooks the edges of the sheets together and taps them down.

   _____

C Phrases refresher

Give safety instructions for working in the welding shop. Use the phrases below and write down complete sentences.

It is necessary to … You should / shouldn't … The … will protect … Don't forget to …

1. Weld the two metal bars with an arc welding technique.

    use welding hood

    _____

    wear fire-resistant gauntlet gloves

    _____

    check wire speed control

    _____

2. Prepare the welding process with a gas welding technique.

    adjust pressure in gas and oxygen cylinders

    _____

    protect your body with leather apron

    _____

    wear welding mask

    _____

3. Solder the two tin sheets together.

    always wear protective clothing

    _____

    touch hot solder seam

    _____

    wait until solder has cooled down

    _____

A Module refresher

1 Complete the sentences with the words / phrases from the box.

> pull • blunt • clamped • the main switch • sharpened • revolution speed • locked in • recharged •
> set • centre punched • centred • turn

1. The twist drill does not cut properly because it is _____ and needs to

   be _____ .

2. The power drill is not working because the battery must be _____ .

3. What is wrong with the column drill? Have you adjusted the required _____?

4. Have you _____ the drill in the chuck properly?

5. Could you please check if the machine is _____ correctly?

6. The end mill has not cut the correct slot. – Maybe the machine vice has not been _____ .

7. If the lathe doesn't start, you must check _____ .

8. Another reason why the machine doesn't start may be that the safety button is _____ .

9. To release the safety button, you must _____ it a bit and then _____ it out.

10. The hole wasn't drilled as marked because it wasn't _____ before drilling.

2 Look at the text from a DIX power tool manual. Point out the main aspects regarding proper drill
applications and the adequate drilling speed. Then write a short text about these aspects in German.

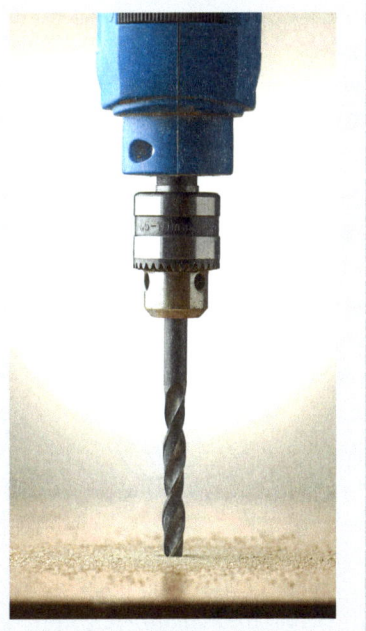

Which drill for which application?

Before using a drill, certain aspects have to be considered. First of all,
it is important to know the properties of the material to be drilled.
Materials such as wood, metal, stone or concrete determine what kind
of drill bit should be used and which drilling speed should be chosen.
When drilling wood, use a twist drill bit with a brad point, which drills
holes with small to medium diameters. A twist drill bit, also called an
HSS (high-speed steel) drill, should be used for drilling metal. When
drilling stone or concrete, a tungsten-carbide drill bit is the best. Quite
often do-it-yourselfers prefer to use one multi-purpose drill that can
cope with most common materials.
Which drilling speed? You can find information about the correct
drilling speed on the packaging or in the operating instructions.
However, the basic rule is: the smaller the drill bit diameter, the higher
the speed, and the larger the drill bit diameter, the lower the speed.
A second rule is: the softer the material, the higher the speed, and the
harder the material, the lower the speed.

3   **Read the following recommendations about the maintenance of a power drill. Find the incorrect information in the sentences and correct it.**

1. The chuck of the machine must be kept clear at all times to prevent overheating.

   _____

2. If a machine is used mainly for drilling wood, bore dust should be removed regularly from the chuck.

   _____

3. To remove the dust, hold the machine vertically with the chuck facing up, and blow the dust off the chuck completely. The collected dust will drop from the chuck.

   _____

4. Regular use of cleaner spray for the drill bits is recommended.

   _____

5. Use original tool accessories and spare parts, if available.

   _____

6. If components which have not been described in the manual need to be replaced, you can buy them in any DIY market.

   _____

4   **Which measurements are taken with the following tools? Write complete sentences.**

Example: *The length and diameter of a bolt or the depth of a hole are measured with the vernier calliper.*

1. thread gauge _____

2. folding rule _____

3. feeler gauges _____

4. steel square _____

5. dial indicator _____

6. micrometer _____

7. gauge blocks _____

B   Grammar refresher

Reported speech

Im Englischen wird die indirekte Rede *angewendet um auszudrücken, was jemand gesagt hat, besonders nach typischen Verben wie* say, tell, complain, promise.
Example: Trainee: "**I'm having trouble** with the power drill type 700 SP."
The trainee told the service engineer that **he was having trouble** with the power drill type 700 SP.
Es wird die Zeitform des Wortes gewählt, das die indirekte Rede *einleitet.*
Example: Helen: "**I'll give** him the drill bits tomorrow."
Helen said (that) she **would give** him the drill bits the following day.
→ **SB p. 230 Reported speech**

1 **Report the following statements.**

1. Instructor to Mike: "Tighten the chuck completely before you start the main spindle." (told)

   _____

2. Instructor to trainee: "Use tungsten-carbide-tipped drill bits for drilling steel." (advised)

   _____

3. Instructor: "And don't forget to switch to percussion drilling before drilling concrete." (added)

   _____
   _____

4. Trainee to another trainee: "When drilling aluminium use turpentine or paraffin." (answered)

   _____

5. Instructor to trainee: "Use a ring spanner of the appropriate size." (told)

   _____

6. Instructor to trainee again: "Centre punch the position of the hole to be drilled in the metal plate before you start drilling." (also advised)

   _____
   _____
   _____

7. Matt: "I'm going to meet the service engineer at the entrance of our shop floor after lunch." (said)

   _____
   _____

8. Steve: "The drill will be repaired tomorrow." (promised)

   _____

9. The secretary on the telephone: "I've been waiting for you to ring back for two hours!" (complained)

   _____
   _____

10. Trainee to instructor: "Sorry, Sir, the component isn't finished yet." (apologised)

    _____

Reported question

Die indirekte Frage _wird angewendet nach typischen Verben, die jemanden auffordern etwas zu tun wie_ request, remind, ask. _Oft wird der Infinitiv verwendet. Wird die_ indirekte Frage _mit einem Fragewort eingeleitet, ist die Wortstellung wie im Aussagesatz. Das die_ indirekte Frage _einleitende Wort bestimmt die Zeitform._
Example: Sean: "Please **hand** me the box with the Allen key bits."
Sean **asked** Bob to **hand** him the box with the Allen key bits.
→ **SB p. 230 Reported speech**

---

2 **Report the following requests and questions.**

1. Sean to Phil and Helen: "Don't forget to check the main cable of the power drill." (reminded)

   _____

2. Trainee to another trainee: "What lubricant should I use when I'm drilling aluminium?" (asked)

   _____

3. Benton to Cliff: "Have you set the switch anticlockwise?" (asked)

_____

4. The instructor to the trainee: "Did you check the quality of the lubricant yesterday?" (asked)

_____

5. Sean: "The machine isn't running and I don't know why." (wondered)

_____

6. Bob to Matt: "Did you clamp the material properly in the three-jaw chuck?" (asked)

_____

7. Steve: "Why doesn't the milling cutter make a proper slot?" (wanted to know)

_____

8. Trainee: "How is the material centred for the turning operation?" (wondered)

_____

9. Matt to Bob: "Check if the milling cutter is sharp enough." (reminded)

_____

10. Cliff to Benton: "Did you forget to release the safety button?" (asked)

_____

C   Phrases refresher

Complete the following telephone conversation with the hotline of a tool company with suitable phrases. Student A starts.

Student A	Student B
Hello, this is TOOLS PERFECT in Manchester. **1** _____?	Hello, this is Benton Thomas from DEP in Stuttgart. We've got **2** _____ with your power drill TOP 800 silent.
What **3** _____?	It's the reversing switch **4** _____.
I see. When **5** _____?	About three months ago. Can we **6** _____?
Yes, of course. We give a **7** _____ on all our products.	Ah, that's fine. What **8** _____?
Just **9** _____ to an authorised service centre. **10** _____ a list of our service centres at the back of your manual.	Oh, fine. Thank you **11** _____.
You're **12** _____. **13** _____.	**14** _____.

Module 7 Properties of materials

A Module refresher

1 List the mechanical properties of materials that are mentioned in the following text.

Before working with certain materials, an engineer has to think of the mechanical properties. Is the material suitable for the planned job? For some jobs hard or dense material is needed. Or the material has to be malleable for example when working with sheet steel, which can be rolled or pressed. However, when working with cutting tools such as milling cutters or lathe tools, hardness is required to cut metal. In other cases hardness is necessary for cutting stone, glass or ceramics. Those cutting tools such as drills or sawing discs are often coated with industrial diamonds.

1. _____

2. _____

3. _____

2 Say which engineering materials the properties in the text refer to and give reasons for your answer.

_____

_____

3 Complete the table by filling in the correct metals and non-metals.

glass • gold • steel • leather • cast iron • aluminium • wood • lead • copper • stone • silver • porcelain • mild steel • plastics • titanium

Metals	Non-metals

4 Sort the materials from the table in exercise 3 into the list below.

ferrous metals:

non-ferrous metals:

natural materials:

artificial materials:

5 **Read the text about alloys and answer the questions.**

Materials are rarely used in their pure forms because they are weak and corrode easily, but mixed with other metals or non-metals, their properties are improved and we get alloys.
Almost all metals can be used to form alloys. Steel, for example, is an alloy of iron and carbon. It is harder than pure iron and more resistant to corrosion. In industry, alloys are much more widely used than pure metals. Pure aluminium, for example, corrodes easily, but when alloyed it is suitable for the construction of aeroplanes, cars, and bicycles. Another alloy is stainless steel. It contains chromium which provides a long life and shiny appearance.
Cast iron alone is brittle. However, when alloyed, its properties include excellent machinability, hardness and wear resistance. The applications of alloyed cast iron range from machine tools to cylinder blocks for the automotive industry.

1. Describe the differences between pure materials and alloys. Give a definition of an alloy.

_____

2. Outline the advantages of stainless steel.

_____

6 **Summarise the main points about cast iron in German.**

_____

_____

7 **Listen to the dialogue about the tensile test and put the different steps into the correct order.**

A 2.13

☐ The shoulders of the specimen are fixed in the tensile tester.

☐ The distance between the gauge marks is measured again.

☐ The tensile strength is measured with a computerized system.

☐ One end of the specimen is pulled away from the other fixed end.

☐ The specimen necks at the fracture point.

☐ The distance between the gauge marks is measured to get the original gauge length.

☐ The maximum stress is achieved.

☐ Gauge marks are made on both sides of the point where the fracture will occur.

8 **Listen to the dialogue again and name the metal properties that are mentioned.**

_____

B Grammar refresher

Gerund

Im Englischen kann man wie im Deutschen ein Verb in ein Substantiv verwandeln. Diese Wortbildung, das Gerundium, wird durch -ing + Infinitiv (Grundform des Verbs) gebildet.
Das Gerundium kann sowohl Subjekt als auch Objekt sein.
Example: **Filing** is rarely done by hand. He began **cleaning** the machines.
Das Gerundium steht nach:
– Präpositionen (Verhältniswörtern) oder präpositionalen Ausdrücken
Example: He **was afraid of starting** the machine. The trainee **apologised for being** late.
– nach bestimmten Verben (Zeitwörtern) z. B. enjoy, like, dislike, remember, stop
Example: He **enjoys working** with his colleagues from England.
→ **SB p. 233 Gerund**

1 Complete the sentences using the phrases from the box and the correct form of the verbs in brackets.

took great pleasure in • were in danger of • apologised for • insisted on • were capable of •
were looking forward to • are interested in

Example: Car builders (?, to learn) about new and lighter materials.

Car builders are interested in learning about new and lighter materials.

1. The manager of the machine tool company (?, to present) the new plant to the public.

   _____

2. The manager (?, to keep) the press waiting for over half an hour.

   _____

3. Then he (?, to take) them personally on a tour through the factory.

   _____

4. The sales department said that they (?, to see) the sales figures rise.

   _____

5. The politicians said they were glad to see that the foreign countries (?, to help) themselves.

   _____

6. The apprentices said: "Twelve months ago we (?, to lose) our jobs. Now we are facing a hopeful future."

   _____

Gerunds and infinitives

remember + gerund = *sich erinnern, etwas getan zu haben*
Example: I remember **repairing** the tool.
remember + infinitive = *daran denken, etwas zu tun*
Example: I have to remember **to buy** a new catalogue.
to stop + gerund = *aufhören mit (z. B. einer Tätigkeit)*
Example: He stopped **working**.
to stop + infinitive = *aufhören (mit etwas anderem), um etwas zu tun*
Example: He stopped (working) **to talk** to me.
→ **SB p. 233 Gerund**

2 Put in the correct verb forms.

1. She remembered clearly that she handed in the documents yesterday.

   She remembered _____ (to hand in) the documents yesterday.

2. Don't you remember that you changed the password a week ago?

   Don't you remember _____ (to change) the password a week ago?

3. She looked up from her computer and stopped _____ (to type) because she wanted to listen to what she was hearing on the radio.

4. No one knows why they stopped _____ (to talk) with each other.

5. They immediately stopped _____ (to prepare) a new job on the CNC machine.

6. The engineer remembered _____ (to reset) the CNC programme.

C Phrases refresher

1 Describe the materials and products. The phrases in the box may help you.

It is a strong / sturdy / hard/ corrosion resistant rigid / soft / light / shiny material. • It is something new / classic / timeless. • It is easy / difficult to work with / can't be painted. • It looks good / modern / elegant / state-of-the-art / old-fashioned. • It lives long / is used as protection gear.

1. Wood is _____

2. Steel is _____

3. Chromium is _____

4. Aluminium is _____

2 Say why plastics are more and more used in industry by filling in the gaps with the words in the box.

rot • light (2x) • corrosion-resistant • preferred • heat-resistant • deform • strong • live long • made of • cheaper

1. Casings for computers and printers are _____ plastics because they are _____ .

2. Engineering plastics are used in the automobile, aerospace and packaging industries because

   they are _____ , _____ and _____ .

3. Plastics are used in industry because their use in production is _____ than other materials.

4. Plastics are _____ because they are _____ and do not _____ .

5. In the automotive industry bumpers, for example are made of plastics because they do

   not _____ as easy as metal bumpers and because they are _____ .

Module 8 Electricity basics

A Module refresher

1 **Listen to a text about the history of primary cells. Then answer the questions.**

A 1.28

1. Of which components does a voltaic pile consist?

   _____

2. How did Volta generate current?

   _____

3. What is a primary cell?

   _____

4. Which problem arises when you use primary cells?

   _____

2 **Describe for which jobs you use the tools below.**

   Example: <u>With a multimeter you can measure current, voltage and resistance.</u> _____

1. cable drum / cable reel: _____
2. insulated electrician's pliers: _____
3. insulated electrician's screwdriver: _____
4. insulating / electrician's tape: _____
5. side cutter: _____
6. voltage tester: _____
7. wire stripper: _____

3 **Complete the German translations.**

1. Installation cables in Great Britain look somewhat different from those in Germany. – Installationsleitungen _____.

2. It is also sheathed cable, but the earth wire is not insulated. – Es handelt sich um _____ _____.

3. The colours of the insulation are the same. – _____ _____ sind gleich.

4. The insulation for the live wire is brown. – Die Isolierung _____.

5. The insulation of the earth wire in flexible cable is green and yellow. – _____ _____ ist grün und gelb.

33

4 Connect the sentence beginnings on the left with the endings on the right.

1. Every time I switch on the	a) so you can easily reset it.
2. When a fuse trips,	b) motor the fuse trips / blows.
3. It is an automatic circuit breaker,	c) with an RCD (residual current device).
4. Not only bathrooms or kitchens must be equipped	d) this can mean that there is a short circuit.
5. My voltmeter only needs batteries	e) you have to connect it in parallel.
6. If you want to measure current,	f) for measuring resistance.
7. If you want to measure the voltage,	g) you have to connect your ammeter in series.

5 Put the jumbled sentences into the correct order.

1. a primary / consists of / and a secondary coil / A transformer

   _____

2. are easy / to handle / Screw-type terminals

   _____

3. step-up / step-down transformers / There are / and

   _____

4. should have / Power supplies / protection built in / an over-voltage protection or an over-current

   _____

5. or a solenoid with an iron core / of an electromagnet / The flow of current through a coil / represents

   the simplest form _____

   _____

6. armature that can open or close / A relay consists / and an iron / when control voltage is applied to

   the coil / electric contacts / of a coil _____

   _____

B Grammar refresher

Past Tense und Present Perfect

Das Past Tense *wird bei regelmäßigen Verben durch Anhängen von -ed an den Wortstamm gebildet. Bei unregelmäßigen Verben verwendet man die zweite Stammform.*
Das Present Perfect *wird bei regelmäßigen Verben mit einer Form von* have / has + verb + ed, *bei unregelmäßigen Verben mit* have / has + *dritter Stammform gebildet. Es drückt aus, dass eine Handlung noch nicht abgeschlossen ist bzw. für die Gegenwart noch von Konsequenz ist.*
Das Past Tense *drückt Handlungen aus, die vergangen und abgeschlossen sind. Es gibt bestimmte* Signalwörter *für diese Zeiten:*
Past Tense: *ago, yesterday, last year, when, Zeitangaben in der Vergangenheit*
Present Perfect: *yet, already, just, since (Zeitpunkt), for (Zeitraum)*
Examples: They **came back** from the customer at eight **last night.**
He **has just informed** me about the damage.
→ **SB p. 224 Past Tense / p. 225 Present Perfect**

Technical Milestones
Englisch für technische Berufe

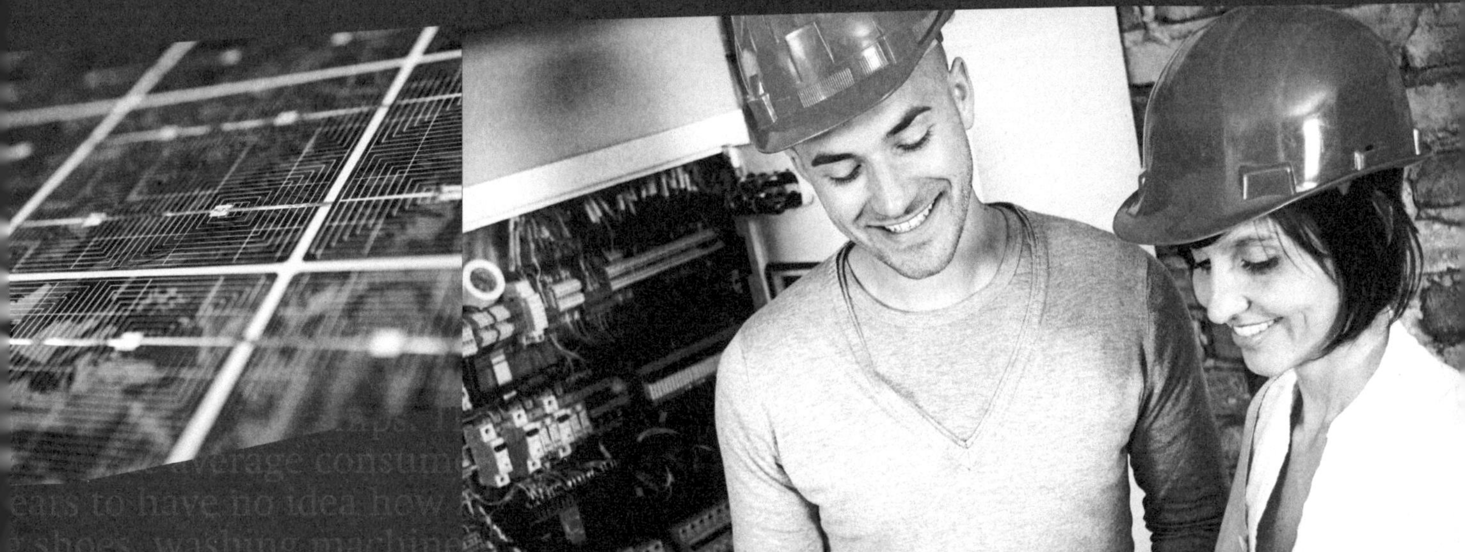

Lösungsheft zum Workbook

Hinweis: Die Audioskripte sind im Anhang des Schülerbuchs abgedruckt. Alle Audioskripte zu zusätzlichen Hörtexten sind in diesem Teil des Workbooks bei den dazugehörigen Aufgaben abgedruckt.

Module 1 Refresher course: International communication S. 5

A Module refresher

1 Complete the following sentences. . . .

A	D	B	N	M	I	T	I	S	A	T	I	I	S	A	R	T
B	U	C	H	A	R	E	S	T	R	H	A	D	O	Q	N	O
E	T	R	I	D	O	V	W	H	E	P	F	I	M	L	D	A
R	C	L	K	R	O	M	E	O	N	O	R	W	A	Y	O	B
I	H	V	T	I	S	A	D	E	K	L	K	A	P	S	H	A
A	G	Y	R	D	P	F	I	N	N	I	S	H	I	N	I	T
R	H	L	K	F	O	W	S	G	H	S	O	U	J	E	K	O
I	O	N	U	E	D	W	H	I	L	H	I	N	T	A	P	F
F	W	J	A	V	B	G	E	Q	M	O	H	G	O	B	H	A
A	W	V	R	K	R	B	A	P	N	K	G	A	H	O	K	F
D	H	G	Y	L	A	N	Q	V	H	E	P	R	A	G	U	E
I	O	F	I	D	E	U	K	I	J	M	L	I	V	M	V	G
Y	O	M	A	D	R	U	B	R	E	A	O	A	S	U	P	Q
A	R	T	O	P	K	M	A	T	I	N	L	N	K	O	M	A

2 You have just arrived at the Romanian . . .
Individuelle Schülerbeiträge.
Lösungsvorschlag:
1. The aisle is the passage or area between two rows of seats in a plane, a church etc. where you can walk.
2. A sample case is a suitcase or a travel bag that contains the items a salesperson shows and hopes to sell to the customers he or she visits.
3. A trade fair is an exhibition that lasts several days where companies present their products to possible buyers or the public.
4. A flight attendant is someone who takes care of passengers' needs and serves meals during a plane trip.

⊙ A 1.2
3 Listen to the dialogue and complete . . .
1. meal; 2. hungry; 3. food; 4. beef; 5. vegetables; 6. restaurants; 7. lunch

4 Complete the sentences. Finish them in . . .
Individuelle Schülerbeiträge.

B Grammar refresher

1 Complete the safety instructions given . . .
1. under; 2. into; 3. above; 4. in; 5. in front of; 6. behind;

7. on; 8. for; 9. from; 10. above; 11. under; 12. out; 13. on; 14. on

2 Complete the sentences by filling in the . . .
1. are; 2. are; 3. He is; 4. we; 5. you; 6. We; 7. be

3 Complete the sentences using much, . . .
1. little; 2. many, few; 3. much, many

4 Complete the sentences using some . . .
1. some; 2. any, some; 3. any, any

5 Put the sentences below into English . . .
1. Oliver was good at maths at school.
2. He worked with the company Hieke CNC Technik in Hanover.
3. He wanted to improve his English skills.
4. He decided to work in England for a year.

C Phrases refresher

Complete the conversation using suitable . . .
1. Excuse me; 2. No problem; 3. I'm; 4. I'm from; 5. My name is; 6. Pleased to meet you; 7. Please call me

Module 2 The new company S. 9

A Module refresher

⊙ A 1.5
1 First read the text below. Then listen to . . .
1. pleasant; 2. although; 3. delay; 4. colleagues; 5. Spain; 6. Pleased; 7. Pleased; 8. too; 9. may; 10. introduce; 11. in charge of; 12. engineering; 13. quality control; 14. welcome; 15. Engines; 16. enjoy; 17. sure; 18. various; 19. brief

⊙ A 1.7
2 Hören Sie sich eine Führung durch die . . .
Emily Miller – kurzer Überblick über die Firma: Produktpalette, Kunden; nimmt uns zur Forschungs- und Entwicklungsabteilung mit
John Smiley – Information über Brennstoffzellen-technologie
Peter Thompson – führt uns in die Laboratorien; hoffentlich sehen wir dort ein laufendes Experiment
Steven Hill – zeigt uns interessante Aspekte der Produktion
Jane Owtrim – klärt uns über die Lagerverwaltung auf
Silke Meier – historische Daten und Pläne für zukünftige Entwicklungen

B Grammar refresher

1 Put in the adjectives in brackets in the . . .
1. faster, more accurate; 2. more complicated; 3. best; 4. pleased; 5. most

2 Complete the presentation at a . . .
1. wonderfully; 2. competitive; 3. sharply; 4. considerably;
5. better; 6. exceptionally well; 7. greatly; 8. sharply;
9. carefully; 10. competitive

C Phrases refresher

1 Read the situations carefully. Then tick . . .
1. 1, 3; 2. 2; 3. 1, 2, und 3; 4. 2; 5. 1, 3; 6. alle möglich

2 Sie arbeiten in der Einkaufsabteilung . . .
Lösungsvorschlag:

> Subject: Meeting next Friday
>
> Dear Mr Silverstone
>
> Thank you for your email of yesterday. Thank you
> as well for the presentations and drawings you
> attached which I could open without any difficulty.
> I hereby confirm our meeting next Friday. Ms
> Metzler, who is our Head of Production, is also
> coming to the meeting. I have prepared the
> documentation and will bring a copy of the new
> contract with me.
> I look forward to our meeting next week.
> Yours sincerely,
>
> . . .

3 Sie haben auf der CeBIT in Hannover . . .
Lösungsvorschlag:

> Dear Sir or Madam
>
> At the CeBIT in Hanover I received a leaflet about
> the new flat screen FlatCad X14. Unfortunately,
> I was not able to get in touch with any of your
> representatives about this product.
> At SoftDesign, the company I work for, we would like
> to know whether this screen is suitable for technical
> drawing programmes. This is important for us as we
> design 3D machine parts and it is necessary to look
> at these parts from different perspectives.
> We need 15 screens and hope that you can make
> a good offer. Could you also let us know about the
> terms of payment and delivery?
> Please send this information to the following email
> address: [insert email address here].
> I would be happy to hear from you soon.
>
> Yours sincerely,
> . . .
> Manager of Purchasing

4 Put in the missing words or phrases . . .
Individuelle Schülerbeiträge.

Module 3
Information Technology S. 13

A Module refresher

1 Complete the sentences with suitable . . .
1. physical; 2. mouse; 3. WLAN; 4. switch; 5. storage;
6. ESC; 7. shift; 8. @; 9. joystick

2 Put the following sentences into the . . .
1. The space bar produces a blank space when you
   press it.
2. With the page down or page up key you can leaf
   through a manuscript pagewise.
3. There was no single inventor of the computer.
4. In the 1830s Charles Babbage invented a mechanical
   calculator.
5. You should control computers but they shouldn't
   control you.

3 Put the sentences below into German.
Lösungsvorschläge:
1. Charles Babbage erfand eine mechanische
   Rechenmaschine in den 30er Jahren des
   19. Jahrhunderts (in den 1830er Jahren). Aber Blaise
   Pascal und Wilhelm Leibniz gehören auch zu den
   Erfindern der mechanischen Rechenmaschinen.
2. Deutsche Computerspezialisten, die etwas über
   die Geschichte der Computer wissen, werden
   Ihnen / dir sagen, das Konrad Zuse den ersten
   programmierbaren Computer in Berlin zwischen
   1936 und 1941 baute.
3. 1971 kam der erste Mikroprozessor auf den Markt.
4. Als Vakuumröhren durch Transistoren ersetzt
   wurden, erhielt die Entwicklung von Computern
   einen neuen Anstoß.
5. Computer kontrollieren zu viel unseres Alltagslebens
   – wenigstens sagen das die Kritiker.
6. Man weiß heute immer noch nicht, in welche
   Richtung sich das Internet entwickeln wird.
7. Problem Nummer 1 wird ohne Zweifel die Sicherheit
   sein.
8. Eine Sache ist definitiv klar: Das Internet wird mehr
   und mehr Aufgaben übernehmen.
9. Der Hauptvorteil von Modell A gegenüber den
   Modellen B und C ist, dass alle Hardwarekomponen-
   ten bei Modell A wirklich auf dem neuesten Stand
   der Technik sind, wobei die anderen zwei Modelle
   nur die Standardausstattung beinhalten.

B Grammar refresher

1 Put the words in the box into the gaps . . .
1. shouldn't; 2. needn't; 3. should; 4. can; 5. mustn't;
6. should; 7. must; 8. need to; 9. must; 10. might;
11. shouldn't; 12. can't

2 Complete the sentences with the . . .
1. Were you able to; 2. was not allowed to; 3. did you
have to do; 4. had to; 5. had to

C Phrases refresher

A customer of yours is looking for a new ...
Individuelle Schülerbeiträge.

Module 4
Mechanical engineering –
Tools S. 17

A Module refresher

1 Look at the lathe, milling machine and ...
centre lathe

1 tailstock; **2** headstock with gearbox; **3** machine bed with slide; **4** emergency stop; **5** saddle and cross slide; **6** tool quick exchange; **7** handwheel control; **8** toolpost; **9** main spindle with chuck; **10** control rod

milling machine

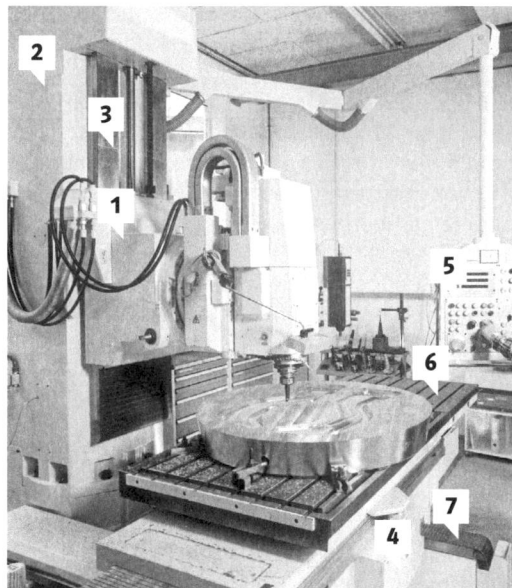

1 overarm; **2** column; **3** vertical slide; **4** knee; **5** control panel; **6** machine table; **7** base

grinding machine

1 table traverse; **2** column; **3** table hand feed; **4** grinding disc; **5** base; **6** machine table; **7** emergency stop; **8** start and stop switch

2 Complete the crossword by filling in ...
Down
1. milling machine; 2. pair of pliers; 4. grinding machine; 5. bench vice; 8. riveting tongs

Across
3. screwdriver; 6. column drill; 7. hammer; 9. spanner; 10. Allen key

⊙ A 2.12
3 Match the tools mentioned in the ...
1. headstock; 2. saddle and cross-slide; 3. machine vice; 4. table traverse; 5. chuck; 6. toolpost

Audioskript:
Brian: Look here, Sven. I've got a brochure with some information about machine tools. They talk about the typical parts of machine tools. For example, a lathe has a base and it is equipped with a saddle, a headstock, a three-jaw chuck and a toolpost. The headstock houses the gears that drive the main spindle with the chuck. The main spindle then rotates the workpiece which is held by the chuck.
Sven: What do they say about the saddle and cross slide?
Brian: Let me see. Yes, the saddle and cross slide feeds the cutting tool. The cutting tool itself is mounted to the toolpost.
Sven: Look what it says here. The rotation of the main spindle can be performed either clockwise or anticlockwise. You just set the motor-reversing switch in the corresponding position.
Brian: That is interesting. Did you pick up a brochure about milling machines?
Sven: Yes I did. Ah, here it is. This brochure has information about milling and grinding machines. A milling machine has three main parts: a column

mounted on a base, an overarm with the spindle and a vertical slide for the machine table. The table supports a machine vice which clamps the workpiece in the position you want it to have.

Brian: That grinding machine on the second page seems to have similar parts to a milling machine. Both machine tools have a base and a column. However, the grinding machine also has a table traverse to feed the material to the grinding wheel. This model has room in the base table to store the grinding wheels.

Sven: That's a good idea. Then the wheels are always ready to use.

B Grammar refresher

1 Ask questions about the words in blue.
1. What did Mr Miller show his visitors?
2. Why do the trainees get basic training on the machine tools?
3. Where can you see a column drill?
4. When will the milling machine be set up?
5. How are the riveting tongs used?
6. Which spindle rotates the three-jaw chuck?
7. When does the machine start running?

2 Use the correct relative clauses to ...
1. The visitors to the company, who are from abroad, are asked not to take photos during the tour.
2. The instructor who is presently setting up a milling machine is preparing a training sequence for his apprentices.
3. The machine vice, which is mounted on the machine table, holds the material in the position required.
4. Alexander needs some help with his milling job, which is described in the work plan.
5. The turner who is highly skilled in machining processes must exchange the reversible carbide tip.
6. The latest lathe model, which was demonstrated to the visitors, is equipped with a high speed spindle.
7. The finishing process of the milled part is prepared by Sven, who is in his last year of training.
8. The firm's visitors liked the demonstration which was about CNC machining processes.

C Phrases refresher

1 Complete the following sentences of a ...
1. latest; 2. feature; 3. benefit; 4. advantage; 5. works; 6. reliable

2 Write down the safety instructions for ...
1. Wear a mask or a handshield.
2. You need to wear safety boots.
3. Watch out for faulty insulation.
4. Don't forget to wear some goggles.

Module 5
Joining and assembly S. 21

A Module refresher

1 Match the different joining methods ...
Mechanical joint
riveting; screwing; fitting; nailing

Non-mechanical joint
welding; soldering; lasing; bonding

2 Look at the illustrations and list the ...
the oxygen and acetylene cylinders; the acetylene and oxygen regulators; the gas and oxygen hoses

3 Write down the tools which are used ...
welding   – welding torch
riveting   – riveting tongs, rivets
bonding   – adhesives
screwing – screwdriver, spanner, Allen key, screws, bolts
fitting   – mallet, bench vice, fitting parts, press fits, folded joints
lasing   – laser gun
soldering – soldering iron
nailing   – hammer, nails

4 Two trainees are discussing various ...
Lösungsvorschläge:
1. A: Do you remember how two base plates of a machine vice are assembled?
   B: Oh yes, the base plates of a machine vice are assembled using bolts and an Allen key.
2. A: How do you repair the broken handle of a tea cup?
   B: It is bonded with adhesive.
3. A: Do you know how flat steel plates are joined to form a square?
   B: That' clear, the flat steel plates are welded to form a square.
4. A: Which tools are used to mount the aluminium shelves in the store room?
   B: The aluminium shelves are mounted with bolts and nuts by using a spanner.
5. A: Do you remember how the copper wire is fixed to the circuit board?
   B: Oh yes, the copper wire is soldered to it.
6. A: How do you hang a picture on the wall?
   B: You drive a nail in the wall with a hammer.
7. A: Do you know how a wooden book shelve is assembled?
   B: That' easy. It is assembled with screws and a screwdriver.

⊙ A 1.17
5 Listen to the dialogue and list the parts ...
1. blue cylinder with oxygen; 2. regulator; 3. gauge; 4. torch; 5. flip-front helmet

6 Match the English and the German ...
1. c); 2. b); 3. d); 4. a)

B Grammar refresher

1 Complete the following sentences by ...
1. are required; 2. are used; 3. are needed; 4. be driven in; 5. are shot; 6. are fastened; 7. is inserted; 8. is mixed; 9. be adjusted; 10. is regulated

2 Describe how a locking seam ...
1. Sheet metal can be joined in different ways.
2. Lines are scribed on the sheet metal where the folding edges are wanted.
3. The first sheet is clamped in the vice.
4. The edges are tapped to a 90-degree angle.
5. Then the edge is bent to almost 180 degrees.
6. The process is repeated with the second sheet of metal.
7. Finally, the edges of the sheets are hooked together and tapped down.

C Phrases refresher

Give safety instructions for working in ...
Lösungsvorschläge:
1. – You should use a welding hood.
   – It is necessary to wear fire-resistant gauntlet gloves.
   – Don't forget to check the wire speed control.
2. – It is necessary to adjust the pressure in the gas and oxygen cylinders before you start.
   – The leather apron will protect your body.
   – Don't forget to wear a welding mask.
3. – It is necessary to always wear protective clothing.
   – You shouldn't touch the hot solder seam.
   – It is necessary to wait until the solder has cooled down.

Module 6
Troubleshooting, maintenance and warranties S. 25

A Module refresher

1 Complete the sentences with the words ...
1. blunt, sharpened; 2. recharged; 3. revolution speed; 4. clamped; 5. set; 6. centred; 7. the main switch; 8. locked in; 9. turn, pull; 10. centre punched

2 Look at the text from a DIX power tool ...
Lösungsvorschlag:
The main aspects of proper drill applications and the adequate drilling speed
You should know the material to be drilled. A twist drill with a brad point is used for drilling wood.
When drilling metal, use a HSS drill bit; for drilling stone or concrete a tungsten, carbide drill is recommended. Information about the adequate drilling speed can be found on the packaging or in the operating instructions. A basic rule is as follows: the smaller the diameter of the drill bit, the higher the drilling speed, and of course the other way round, i.e. the larger the drill bit diameter, the lower the speed.

Another rule is about the material in relation to the drilling speed. For example, softer material requires higher speed, and harder material requires lower speed.

Mediation
Der zu bohrende Werkstoff muss bekannt sein. Für Holz wird ein Spiralbohrer mit Zentrierspitze verwendet. Für Metall nimmt man einen HSS Bohrer, für Stein oder Beton eignet sich ein Hartmetallbohrer / wird ein Hartmetallbohrer empfohlen.
Bezüglich der geeigneten Bohrgeschwindigkeit gibt es Informationen auf der Rückseite der Verpackung oder in der Gebrauchsanweisung. Eine Faustregel besagt, je kleiner der Bohrdurchmesser, desto höher die Bohrgeschwindigkeit und umgekehrt, d. h. je größer der Bohrdurchmesser, desto geringer die Bohrgeschwindigkeit. Im Verhältnis zum Werkstoff, in den gebohrt wird, heißt es, je weicher der Werkstoff, desto höher die Geschwindigkeit, je härter der Werkstoff, desto geringer die Geschwindigkeit.

3 Read the following recommendations ...
1. The ventilation slots of a machine must be kept clear at all times to prevent overheating.
2. If a machine is used mainly for impact drilling, collected dust should be removed regularly from the chuck.
3. To remove the dust, hold the machine vertically with the chuck facing down and open and close the chuck completely.
4. Regular use of cleaner spray for the clamping jaws and the clamping jaw borings is recommended.
5. Use only original tool accessories and spare parts.
6. If components which have not been described need to be replaced, contact one of the tool service agents from the list of guarantee / service addresses.

4 Which measurements are taken with ...
1. The thread or thread pitch of a bolt is checked with a thread gauge.
2. A carpenter, for example, measures the length of a wooden board or beam with a folding rule.
3. They are used to check the distance between the electrodes of a spark plug.
4. With a steel square we measure the right angle of a metal square plate.
5. The dial indicator is used to measure the required position of a machine vice for milling slots.
6. With the micrometer we measure the diameter of a bolt, for example, to the accuracy of a hundredth of a millimetre.
7. Gauge blocks are used to check whether the diameter of a drilled hole is good or not.

B Grammar refresher

1 Report the following statements.
1. The instructor told Mike to tighten the chuck completely before he started the main spindle.
2. The instructor advised the trainee to use tungsten-carbide-tipped drills for drilling steel.

3. The instructor added that she/he should not forget to switch to percussion drilling before she/he started drilling concrete.
4. The trainee answered that she/he should use turpentine or paraffin when drilling aluminium.
5. The instructor told the trainee to use a ring spanner of the appropriate size.
6. He also advised her/him to centre punch the position of the hole to be drilled in the metal plate before she/he started drilling.
7. Matt said (that) he was going to meet the service engineer at the entrance of their shop floor after lunch.
8. Steve promised that the drill would be repaired the next day.
9. The secretary complained on the telephone that she had been waiting for the caller to ring back for two hours.
10. The trainee apologised to the instructor that he had not finished the component yet.

2 Report the following requests and ...

1. Sean reminded Phil and Helen not to forget to check the main cable of the power drill.
2. The trainee asked another trainee what lubricants he should use when drilling aluminium.
3. Benton asked Cliff if he had set the switch anticlockwise.
4. The instructor asked the trainee whether he had checked the quality of the lubricant the day before.
5. Sean wondered why the machine wasn't running and he didn't know why.
6. Bob asked Matt if he had clamped the material properly in the three-jaw chuck.
7. Steve wanted to know why the milling cutter didn't make a proper slot.
8. The trainee wondered how the material was centred for the turning operation.
9. Matt reminded Bob to check if the milling cutter was sharp enough.
10. Cliff asked Benton if he forgot to release the safety button.

C Phrase refresher

Complete the following telephone ...

1. Can I help you; 2. some trouble; 3. seems to be the problem; 4. which is blocked; 5. did you buy the tool; 6. make claim under the warranty; 7. 1-year warranty; 8. do I have to do then; 9. send the tool; 10. You'll find; 11. for your help; 12. Welcome; 13. Goodbye; 14. Goodbye

Module 7
Properties of materials S. 29

A Module refresher

1 List the mechanical properties of ...
1. hardness; 2. density; 3. malleability

2 Say which engineering materials the ...
Sheet steel is malleable to be pressed or rolled.
Cutting tools are hard to cut metal, stone, glass, ceramics.

3 Complete the table by filling in the ...
Metals:
steel; cast iron; lead; aluminium; gold; silver; mild steel; titanium; copper

Non-metals:
wood; stone; glass; porcelain; plastics; leather

4 Sort the materials from the table in ...
ferrous metals: steel, cast iron, mild steel, titanium
non-ferrous metals: lead, aluminium, copper
natural materials: wood, stone, leather
artificial materials: glass, porcelain, plastics

5 Read the text about alloys and answer ...
1. **Describe the differences between pure materials and alloys. Give a definition of an alloy.**
   Alloys are mixtures of two or more metals. In this form their properties are better than pure materials and they are harder, live longer and do not corrode easily.
2. **Outline the advantages of stainless steel.**
   Stainless steel lives long and looks shiny due to its chromium content.

6 Summarise the main points about cast ...
Lösungsvorschlag:
Only in the form of an alloy cast iron is easily machinable, hard and resistant. It can be applied for machine tools, cylinder blocks or in the automotive industry.

⊙ A 2.13
7 Listen to the dialogue about the tensile ...
1. Gauge marks are made on both sides of the point where the fracture will occur.
2. The distance between the gauge marks is measured to get the original gauge length.
3. The shoulders of the specimen are fixed in the tensile tester.
4. One end of the specimen is pulled away from the other fixed end.
5. The maximum stress is achieved.
6. The specimen necks at the fracture point.
7. The distance between the gauge marks is measured again.
8. The tensile strength is measured with a computerized system.

Audioskript:
Luke: Hello Nina, hi Carsten.
Nina: Hello Luke. How are you?
Luke: Fine thanks. I am just having some trouble writing up my notes about the properties of metals.
Nina: Yes, we were just talking about that. The details were confusing. Just think about steel. Of course it's hard, but I didn't know that it's tensile, too.

Carsten: And it's even elastic to a certain extent.
Luke: If you think about it, it makes sense to test the properties of a material before you use it for a specific purpose, doesn't it?
Nina: Yes, it does. If you want to build a bridge or an aeroplane, the metal you use needs good tensile strength.
Luke: Tensile strength is also an important property when you build cars. In a crash situation you don't want the metal in the car body to fail immediately.
Nina: Okay, so these potential forces are tested before the sheet steel is used in car production. That's the function of the tensile test.
Carsten: All materials including metals have a ductility limit. And the tensile test tells you how far a material can be stretched before the ductility limit is reached.
Luke: But how is the tensile test carried out?
Nina: Well, this is what I noted last week: You put two gauge marks on the specimen, one on either side of where it will break in the test. Then you get the original gauge length by measuring between the two marks.
Carsten: When do you put the specimen in the tester?
Luke: That's the next step, I think. Then you start the test. One end is pulled until the specimen breaks at the fracture point.
Nina: Yes, and before it breaks, the metal achieves the maximum stress.
Carsten: And after the specimen breaks, you measure the distance between the gauge marks again.
Nina: Well, modern testers are computerized so the computer measures the distance for you. It can work out the percentage of elongation of the metal specimen.
Luke: Now my notes make more sense. Thanks!

8 Listen to the dialogue again and name ...
hardness; elasticity; ductility; tensile strength

B Grammar refresher

1 Complete the sentences using the ...
Lösungsvorschlag:
1. took great pleasure in presenting; 2. apologised for keeping; 3. insisted on taking; 4. were looking forward to seeing; 5. were capable of helping; 6. were in danger of losing

2 Put in the correct verb forms.
1. handing in; 2. changing; 3. typing; 4. talking; 5. to prepare; 6. to reset

C Phrases refresher

1 Describe the materials and products ...
Lösungsvorschläge:
1. Wood is a classic material. It is easy to work with; it is repairable. It can be strong, but it is also burnable.
2. Steel is a hard material; much more durable than

wood. It can be formable, but it is difficult to operate on a machine tool.
3. Chromium is also very hard, it looks shiny, it is corrosion resistant and lives long.
4. Aluminium is a light material. It is corrosion resistant but can't be painted.

2 Say why plastics are more and more ...
1. made of, light; 2. strong, heat-resistant, live long; 3. cheaper; 4. preferred, corrosion-resistant, rot; 5. deform, light

Module 8
Electricity Basics S. 33

A Module refresher

⊙ A 1.28
1 Listen to a text about the history of ...
1. It consists of a stack of metal discs that are divided by a saltwater cloth.
2. He connected wires to both ends of the pile.
3. It consists of a zinc case containing electrolyte in a paste form and carbon red.
4. They cannot be recharged and you have to throw them away once they are empty.

2 Describe for which jobs you use the ...
1. If you are too far away from an outlet, you can use it as an extension. You can also plug in more than one appliance.
2. With electrician's pliers you can grip things / get a hold of things or wires.
3. Electrician's screwdrivers protect you when you are working on energised parts (which you shouldn't do anyway).
4. With this kind of tape you can insulate wires or metal parts.
5. A side cutter is used to cut wire.
6. You use it to test if the wire is live.
7. The wire stripper helps you to strip off insulation material from a wire.

3 Complete the German translations.
1. sehen in England etwas anders aus als die in Deutschland; 2. ummanteltes Kabel, aber die Erdung ist nicht isoliert; 3. Die Farben der Isolierung; 4. der Phase ist braun; 5. Die Farbe der Isolierung der Erdung bei flexiblem Kabel

4 Connect the sentence beginnings on ...
1. b); 2. d); 3. a); 4. c); 5. f); 6. g); 7. e)

5 Put the jumbled sentences into the ...
1. A transformer consists of a primary and a secondary coil.
2. Screw-type terminals are easy to handle.
3. There are step-up and step-down transformers.
4. Power supplies should have an over-voltage protection or an over-current protection built in.

5. The flow of current through a coil or a solenoid with an iron core represents the simplest form of an electromagnet.
6. A relay consists of a coil and an iron armature that can open or close electric contacts when control voltage is applied to the coil.

B Grammar refresher

1 Put the words into the correct order …
1. We talked to the customer yesterday.
2. They repaired the machine a week ago.
3. I've already ordered a multimeter.
4. We sent your consignment on June 14th.
5. I've just received your mail.
6. When did you send the order?
7. I read the manual after he suggested it.

2 Put the following sentences into …
1. We haven't received an order yet.
2. We noticed the problems with the transformer yesterday.
3. The instructions our instructor gave us some time ago were very helpful.
4. We learnt many new functions of the measuring instrument in the training last week.
5. He has inserted the diode wrongly.
6. He hasn't noticed yet that the circuit isn't working.

C Phrases refresher

1 You want to give instructions to some …
Lösungsvorschläge:
1. it is forbidden to; 2. remember to; 3. make sure to; 4. it is important to; 5. be careful with; 6. be sure to; 7. the best thing you can do is to; 8. don't forget to

2 Put the following sentences into English.
Lösungsvorschläge:
1. Remember to put on rubber gloves.
2. It is important to wear safety boots at all times.
3. Be careful with every kind of humidity.
4. It is important to know all the regulations.
5. It is best to talk about difficult repairs with an experienced colleague.
6. Make sure that you know the emergency measures.

Module 9
Energy and the environment
S. 37

A Module refresher

1 Match the sentence beginnings on the …
1. f); 2. d); 3. b); 4. e); 5. c); 6. a)

2 Match the sentences with the six …
A Solar power: Photovoltaic panels convert light directly into electricity. When light shines on silicon layers, an electric current is produced. A single cell produces only very little current.

B Hydroelectric power: Water is stored behind a dam. Potential gravitational energy can be transformed into other forms of energy. The water arrives at the turbines at very high pressure because of its great height.

C Tidal power: A dam is built across the mouth of an estuary. The ebb and flow of the tides can be used to turn a turbine. The difference between low and high tide has to be large enough.

D Wave power: A large area of sea is covered with floats connected together. The kinetic energy caused by up-and-down movements is converted into electricity as in a dynamo.

E Wind power: Huge blades are mounted on top of slim towers. The spinning movement turns a generator, which produces electricity. It has a reputation for making a swooshing noise day and night.

F Nuclear power: It generates huge amounts of electricity from small amounts of fuel. The nucleus is split roughly in half and releases energy in the form of heat. Tiny amounts of waste can endanger all forms of life on earth.

3 Read the text on solar power and …
1. friendly; 2. converted; 3. side; 4. devices; 5. directly; 6. connected; 7. shines; 8. current; 9. individual; 10. generate

◉ A 1.32
4 Listen to Dr Wall's predictions about …
1. He believes that we will solve the energy problems we are facing at the moment.
2. Fuel-cell driven cars will be on the roads within 10 years.
3. Petrol-driven cars will be a thing of the past within 50 years.
4. We will reduce it by using solar-cell technology to its full extent.

B Grammar refresher

Put the verbs in brackets into the present …
1. broken; 2. transporting; 3. doing; 4. found; 5. developing; 6. talking; 7. checked; 8. waiting; 9. looking; 10. Having

C Phrases refresher

1 Put in the most suitable words from …
1. about; 2. on; 3. on; 4. feel; 5. seems; 6. firmly; 7. no doubt; 8. due to; 9. with; 10. up to; 11 really can't; 12. up

2 Professor Winter and his team are …
Lösungsvorschläge:
– From 1860 to 1870 the temperature dropped sharply from 25 to 15 degrees Celsius.
– From 1890 to 1900 the temperature remained constant at 25 degrees Celsius.
– Temperatures fell dramatically from 1910 to 1920 by 17 degrees Celsius.

- In 1930 temperatures reached a low point of 5 degrees Celsius.
- From 1960 to 2020 there was and will continue to be a considerable increase in temperatures.
- The temperature will reach a high of 40 degrees in 2020.
- After 2020 the temperature will drop sharply.

Module 10
Business Trips S. 41

A Module refresher

1 You are preparing for a business trip. ...
1. Make a list of the appointments you have in the city you are visiting.
2. Make sure that your passport / identity card is up to date.
3. Call your travel agent to book your flights / book your flights on the internet.
4. Register with ESTA if you are travelling to the USA.
5. Make hotel reservations.
6. Make arrangements for getting to and from the airport in both the city you are leaving from and the one you are visiting.
7. Find out what the weather will be like where you are going and plan to pack accordingly.
8. Check out what sights or events you might visit if you have free time in the city you are visiting.
9. Buy a small guide- / phrasebook about your destination and the language spoken there.

⊚ A 2.2
2 Listen to the dialogue and complete it.
1. passport; 2. visa regulations; 3. fewer than; 4. Visa Waiver; 5. Excuse me; 6. digital photo; 7. biometric; 8. valid; 9. port of entry; 10. authorisation; 11. payable; 12. expires; 13. carrier; 14. boarding; 15. application

3 Match the sentences or phrases on the ...
1. d); 2. a); 3. e); 4. c); 5. f); 6. b)

B Grammar refresher

1 Complete the sentences below using ...
1. will have; 2. would / 'd have used; 3. will not / won't get; 4. happens; 5. would not / wouldn't be; 6. would / 'd take you out; 7. rains; 8. would / 'd have solved; 9. were

2 Put the following sentences into English.
1. If I hadn't booked this flight, I would / 'd have been too late for the conference.
2. If we get this contract, they will / 'll give us a bonus!
3. What will you do if you are not / aren't able to speak with the manager / director?
4. The hotel would not / wouldn't have cost so much if you had booked a double room.
5. If the weather improves, more people will visit the trade fair.

C Phrases refresher

1 What might the passenger (p) and ...
1. p; 2. t; 3. t; 4. p; 5. t; 6. t; 7. p; 8. p; 9. t; 10. p

2 Complete the following dialogue about ...
1. book; 2. shower; 3. nights; 4. will / 'll arrive; 5. will / 'll depart; 6. single; 7. double; 8. double; 9. non-smoking; 10. rate; 11. breakfast; 12. early; 13. check in; 14. check out; 15. limousine / shuttle service; 16. airport; 17. limousine / shuttle service ; 18. confirm; 19. credit card

3 Put the following phrases in the right ...
1. Please be seated. I'll bring you water right away.
2. Here is the menu.
3. What would you like to drink with your meal?
4. Would you like to order an hors d'oeuvre?
5. Will you be having soup?
6. How would you like your roast beef done?
7. Would you like a salad with your entrée?
8. Can I bring you the dessert menu?
9. How about coffee or tea?
10. It was a pleasure to serve you, sir.

Module 11
Finding a job in the global village S. 45

A Module refresher

1 Read the text below. Then explain the ...
Individuelle Schülerbeiträge. Folgende Punkte sollten miteinbezogen werden:
- Verträge über Produktion / Dienstleistungen ins Ausland auslagern
- Software-Entwicklung
- keine Sozialleistungen
- Vorschriften, Steuer, Gewerkschaften umgehen
- Nordhalbkugel
- Zeitarbeit
- positive Entwicklung – negative Auswirkungen
- Technologie, Industrie
- Industrialisierung, Deindustrialisierung
- Dritte Welt
- rapides Städtewachstum
- landwirtschaftlich geprägte Gesellschaften
- gesundheitsbezogene und soziale Dienste
- Migration aus ärmeren Ländern
- Unzufriedenheit, Gegenbewegung
- schlechte Qualität, Verzögerung
- Kundenkontaktzentrum
- Fremdsprache, fremde Kultur

2 Find synonyms in the text for the ...
1. to avoid; 2. to handle; 3. has proven; 4. accelerated; 5. irate; 6. less prosperous

⊙ A 2.5

3 Listen to the interview. Answer the ...

1. It is the language used in the company and it is essential for the job.
2. He is applying for the job of web designer.
3. He studied information technology and concentrated on networks, or WLANs.
4. He has designed websites for small businesses, including one for an importer of leisure footwear.
5. No, he may work from home.
6. He is expected to work at least 38 hours per week and to log in and out on the company server.

B Grammar Refresher

1 Complete the sentences using the most ...

1. will panic; 2. is arriving; 3. opens; 4. isn't going to sign; 5. are going to build; 6. starts; 7. are going to explore; 8. will react

2 Put the sentences into English using ...

1. Should we have children, I'll stop working.
2. Beginning this fall / autumn he will work at home for the publisher as a translator.
3. Our train will be late because of the strike.
4. I don't think that they are going to accept our offer.
5. They are going to discuss the new advertising campaign on Tuesday.
6. I'm going to defend my point of view at the next meeting.

C Phrases Refresher

1 Join the phrases from the left with the ...

Individuelle Schülerbeiträge.

2 Complete the discussion using the ...

Lösungsvorschläge:
1. Looking at the statistics; 2. This proves that; 3. I think you'll find that; 4. In my experience; 5. As I see it; 6. I'm afraid I don't agree; 7. I see your point, but; 8. In my opinion

3 Read the advertisement and write a ...

Individuelle Schülerbeiträge (vgl. SB S. 157 Nr. 3.1 c).

Module 12
Design for manufacturing S. 49

A Module refresher

1 Sitcomfy presents itself in the company ...

1. locations ; 2. develop; 3. required; 4. seating preferences; 5. adjustable; 6. demanding; 7. mission; 8. exceed ; 9. ventilation; 10. fold-flat; 11. fold-and-tumble; 12. storage

⊙ A 2.7

2 Listen to the dialogue and decide ...

1. False; 2. True; 3. False; 4. False; 5. True; 6. True; 7. True

3 Underline the correct word or phrase of ...

1. report; 2. modifications; 3. according to; 4. comply with; 5. conform; 6. manufacturing; 7. mounted; 8. experimented; 9. flaws; 10. mock-ups; 11. setback; 12. deadline

4 Complete the production process of a ...

1. sheet metal, deep-drawn; 2. tack-welded;
3. body-shells, conveyed, dipped, immersion, booths;
4. sub-assembly; 5. marriage, chassis; 6. fluids

B Grammar refresher

Ask questions about the phrases in ...

1. Where does Manfred call from every Friday?
2. What time are you / we meeting John?
3. Who did he give the report to?
4. Why didn't Mike install the bracket?
5. How much have they invested up to now?
6. Who produced the units?
7. What did Grace send when she received the confirmation?
8. Why isn't the product going to make a profit?
9. How long are they going to need for the job?
10. What would we do if we had more time?

C Phrases refresher

1 Match the sentence beginnings on the ...

1. h); 2. g); 3. d); 4. b); 5. f); 6. c); 7. e); 8. a)

2 Match the phrases with the four ...

Beginning a meeting
o); h); n); c)

Outlining the agenda
i); j)

Running a meeting
d); k); a); b); p); q); e); l); m)

Finishing a meeting
g); f)

3 Put the phrases / sentences into German.

1. Die Firma ist in verschiedene Abteilung gegliedert.
2. Die Leiterin / der Leiter einer Firma wird Geschäftsführer, Generaldirektor oder geschäftsführender Direktor genannt.
3. Jede Abteilung ist in Teams aufgeteilt, die von Team- oder Gruppenleitern geleitet werden.
4. Die Leiterin / der Leiter einer Abteilung wird Abteilungsleiter / in genannt.

Module 13
Control Technology S. 53

A Module refresher

1 Match the devices with the definitions ...
1. proximity sensor; 2. actuator; 3. limit switch; 4. switch;
5. lobe pump; 6. valve; 7. pressure switch; 8. solenoid

2 Frank Walker explains to his apprentices ...
1. magnetic field; 2. without a mechanical switch;
3. voltage is applied; 4. resulting current; 5. pulls a metal
switch; 6. contacts touch; 7. when the coil is energised;
8. when the input coil; 9. relays; 10. logical control

⊙ A 2.10
3 Listen to a recommendation of a new ...
1. False; 2. True; 3. True; 4. True; 5. False; 6. False; 7. True;
8. True

4 Put the phrases on negotiating ...
1. d); 2. g); 3. b); 4. a); 5. f); 6. c); 7. e)

⊙ A 2.11
5 Listen to the dialogue and complete it.
1. key people; 2. small-scale orders; 3. compromise on;
4. impressed; 5. quite competitive; 6. going to be an
issue; 7. reckon we could manage; 8. We might even be
able to; 9. meet our delivery deadline; 10. drafted in;
11. Would you be willing to; 12. subject to; 13. What if we
were to; 14. stretch ourselves; 15. good prospects

B Grammar refresher

Choose the correct terms related to ...
1. impossible; 2. can't possibly; 3. possible; 4. may well;
5. must have been; 6. highly unlikely; 7. may have been;
8. likely to; 9. will probably; 10. certainly won't

C Phrases refresher

Compare the valves using the phrases ...
Individuelle Schülerbeiträge.

KMK-Prüfungsvorbereitung 1 –
Mechanical engineering S. 59

⊙ A 2.14
Aufgabe 1 Rezeption – Hörverstehen   20 VP

Hören Sie sich den Dialog zweimal an und ...
1. motor; 2. gear and control units; 3. main spindle;
4. headstock, the main spindle; 5. anticlockwise;
6. revolution; 7. slideways; 8. saddle and cross slide;
9. chuck

Audioskript:
Volker: Hi Philip.
Philip: Hi Volker. What are you doing?

Volker: I have to do a job on the centre lathe. Can you
help me? How do I start the headstock again?
Philip: You can switch on the motor here. Then the
power is transmitted to the headstock or the
spindles.
Volker: I see. The headstock drives the main spindle
which rotates the workpiece, right?
Philip: That's it. And you can rotate the workpiece
either clockwise or anticlockwise.
Volker: Oh, OK. I want to set the revolution speed of
the spindle. Where do I ... ?
Philip: Here is the speed selection lever. You can use
that.
Volker: Thanks, I remember now. What is the function
of the slideways again?
Philip: Normally the slideways are used to control
the feeds of either the cutting tool or the
workpiece.
Volker: The slideways have a different function to the
saddle and cross slide then.
Philip: Yes, as far as I can remember the saddle
and cross slide feeds the cutting tool to the
workpiece, which is clamped in the chuck.
Volker: That's quite complicated. Can we go through
the process again?

Aufgabe 2 Rezeption – Leseverstehen   20 VP

Lesen Sie den folgenden Text und ...
1. Das Material muss in einen Schraubstock gespannt
werden um einen festen Halt zu garantieren.
2. Die Metallplatte kann entweder mit einer Bügelsäge,
einem Winkelschleifer oder einer Kreissäge
zurechtgeschnitten werden.
3. Die Verwendung eines Winkelschleifers erfordert
weniger Körperkraft.
4. Mit einem Winkel werden die Ecken der Grundplatte
auf Rechtwinkligkeit geprüft.
5. Sie werden mit einer Feile entgratet, zuerst mit einer
Schubfeile, dann mit einer Schlichtfeile.
6. Ein Elektrobohrer wird mit einem Spiralbohrer
verwendet.
7. Die Stellen für die zu bohrenden Löcher werden
zuerst mit einer Anreißnadel markiert und dann mit
einem Körner zentriert.

Aufgabe 3 Mediation   30 VP

Die Auszubildenden der englischen ...
*Es werden 6 – 7 Schutz- und Sicherheitsausrüstungen
erwartet mit der Beschreibung ihrer Funktionen. Pro
Aspekt sind 5 VP vorgesehen.*

Lösungsvorschlag:
"Before we enter the workshop of our shopfloor, we
always put on overalls and safety boots to protect our
feet from falling metal stock.
In the general workshop we do our training at the bench
vice, for example, file metal bars or grind metal. It is
important to wear goggles to protect our eyes. When
we do a drilling job at the column drill, I wear a hairnet

because my hair is rather long. In the welding shop we wear gloves to protect our hands from sharp metal edges and the heat produced during welding operations. We also wear welding masks or use handshields to protect our eyes from the bright light and flying sparks."

Aufgabe 4 Produktion 30 VP

Zusammen mit Ihrem englischen ...
Es werden 5 Arbeitsprozesse erwartet. Pro Arbeitsprozess sind 6 VP vorgesehen.

Lösungsvorschlag:
- To get the raw material to be machined, a bench saw is used to cut the different workpieces.
- The flat and squared components are milled, the holes are drilled and countersunk on a column drill.
- The grooves in the base plate and the jaws are also milled.
- The spindle is turned on a lathe.
- The thread of the spindle is either cut with a thread cutter on a lathe or rolled with a thread roll on a lathe.

Aufgabe 5 Interaktion 30 VP

Sie arbeiten stundenweise in einem ...
Lösungsvorschlag:

Sprecher A:	Good afternoon. How may I help you?
Sprecher B:	I'm looking for a drill.
Sprecher A:	What do you need the drill for?
Sprecher B:	I'd like to do a few jobs in my new flat.
Sprecher A:	I'd recommend a cordless hammer drill. Here's our latest model.
Sprecher B:	What are the special features of this drill?
Sprecher A:	Well, it has a powerful motor with an 18 V-LI battery cartridge and a compact angle gear.
Sprecher B:	I see. Does this model also have a clockwise / anticlockwise rotation?
Sprecher A:	Yes, of course. Furthermore, the drill has a pneumatic switching mechanism.
Sprecher B:	Oh, that's good. And what about the tool change? Does it happen quickly?
Sprecher A:	The new model has SDS-plus chucks.
Sprecher B:	Right. Can I also drill holes with a diameter of 10 mm?
Sprecher A:	Yes, you can even bore holes up to 12 mm with optimal speed under load.
Sprecher B:	Oh, good, that worked really well with the previous model. Could you show me any other models?

KMK-Prüfungsvorbereitung 2 – Welding engineering S. 61

⊙ A 2.15

Aufgabe 1 Rezeption – Hörverstehen 15 VP

Bernd und Oli, zwei deutsche ...
1. Beim Autogenschweißen wird mit einem Sauerstoff-gasgemisch gearbeitet, beim Lichtbogenschweißen mit elektrischem Strom.
2. Der Zustrom von Azetylen und Sauerstoff wird durch Auslassventile geregelt, die sich auf der Gas- bzw. Sauerstoffflasche befinden.
3. Die Ausrüstung zum Autogenschweißen besteht aus einer Sauerstoffflasche und einer Azetylengasflasche. Beide Flaschen sind mit Auslassventilen versehen und jeweils zwei Reglern zur Steuerung des Flaschen- und Betriebsdrucks. Der jeweilige Steuerungsdruck wird an den entsprechenden Messgeräten abge-lesen, die sich an jeder Gasflasche befinden. Jede Gasflasche ist verbunden mit einem Schlauch. Beide Schläuche sind am Schweißbrenner montiert, an dem sich ebenfalls zwei Regler befinden um die korrekte Betriebsflamme einzustellen.
4. Gebraucht werden ein mobiles Stromaggregat, eine Kontaktierungsklemme, eine Erdungsklemme und der Schweiß- oder Fließdraht.

Audioskript:

Amélie:	You know, I haven't done any welding yet in my training programme. I have no idea how all the welding equipment works.
Oli:	No, me neither. Can you start from the beginning, Herr Staiger?
Mr Staiger:	Well, it's difficult to explain welding in a short time. We will start with the equipment today. Does anyone know what this welding gear is called? Bernd?
Bernd:	I don't know the word in English.
Mr Staiger:	This is oxyacetylene welding gear. Both oxygen and acetylene are used in the welding process.
Oli:	The red cylinder contains oxygen. That's mixed with acetylene from the grey cylinder to increase the temperature of the flame. That part on top of each cylinder is a valve and beside it is the regulator with two gauges for each cylinder.
Bernd:	Yes, one gauge is for the pressure in the cylinder and the other is for the working pressure of the torch.
Mr Staiger:	That's right. The flame comes here out of the torch. You adjust the flame to reach the right temperature. Sometimes you need it to be more than 3,000 degrees Celsius.
Amélie:	Can we try out the gear?
Mr Staiger:	Of course, that's why we're here. Bernd and Oli, would you please get the flip-front helmets over there? We'll need them in a minute.

Amélie:	Sorry, Herr Staiger. I have a question. Is electric arc welding gear similar to gas welding gear?
Mr Staiger:	We'll try out the arc welding gear next week but to answer your question, no, we need different equipment for electric arc welding.
Bernd:	What's the difference?
Mr Staiger:	Well, look at the arc welding equipment here.
Amélie:	I see. It runs on electric power. It looks rather small, doesn't it?
Oli:	Yes, it's easy to carry so it can be used in different places to weld thin metals such as car body parts.
Mr Staiger:	That's right, Oli. Here you see the contact tips and the ground clamp. The clamp is clamped to the material while you are welding.
Bernd:	And what are these buttons on the power station?
Mr Staiger:	One button controls the wire speed and the other button sets four different heat positions.
Oli:	Sounds interesting. I'm looking forward to hearing more about it next week.

Aufgabe 2 Rezeption – Leseverstehen   15 VP

Ihre englische Partnerfirma Turntec Ltd. ...

1. Die Schutzbrille sollte Sicherheitsgläser mit Seitenschutz haben.
2. Um die Augen vor herumfliegenden heißen Metallspänen zu schützen
3. Lange weite Ärmel können sich in dem rotierenden Werkstück verfangen und Finger oder Hände einziehen. Langes Haar verfängt sich und kann durch Einzug zu schlimmen Kopfverletzungen führen.
4. Die Schuhe sollten Sicherheitsschuhe aus Leder sein um vor herabfallenden schweren Werkstücken oder Werkzeugen oder scharfen Metallspänen am Boden zu schützen.
5. Der Schmuck kann sich im rotierenden Werkstück verfangen und zu schlimmen Verletzungen an Händen und Fingern führen.
6. Zu beachten ist, dass das Werkstück fest eingespannt ist. Die Drehgeschwindigkeit der Arbeitsspindel sollte langsam angefahren und dann allmählich erhöht werden.
7. Indem man nie vergisst, den Spannschlüssel abzuziehen, ehe die Maschine gestartet wird.
8. Man sollte vermeiden, bei laufender Maschine über das sich drehende Spannfutter zu greifen.

Aufgabe 3 Mediation   30 VP

Ihre Firma hat ein Fax von Mike Masters ...

Lösungsvorschlag:
Zuerst muss man wissen, wozu das Material verwendet werden soll, denn im Maschinenbau ist es wichtig, dass das Material während der maschinellen Bearbeitung nicht bricht. Deshalb muss die Festigkeit z. B. des Metalls bestimmt werden, was durch eine Schlagprüfung, z. B. mit einem Pendelhammer durchgeführt werden kann. Soll die Zugfestigkeit des Materials geprüft werden, wird eine Zugfestigkeitsprüfung vorgenommen, bei der die Zugspannung gemessen wird, bis das Material bricht. Am häufigsten wird in der Industrie die Rockwell-Härteprüfung angewendet. Materialhärte wird definiert als Widerstand gegen Vertiefungen, also wird bei dieser Prüfung die Einkerbung gemessen, die ein Stempel unter bestimmten Druck im Material verursacht. Natürlich werden bei allen Prüfverfahren nur Proben verwendet, die die gleichen Materialeigenschaften haben wie die Bauteile, die maschinell bearbeitet werden sollen.

Aufgabe 4 Produktion   40 VP

Für einen Auftrag möchten Sie bei der ...

Lösungsvorschlag:

> Dear Sir or Madam
>
> We see from your advertisement that you are a supplier of welding equipment.
> Our firm Schweißtechnik GmbH, which is based in Kiel, Germany, has specialised in underwater welding. We are especially interested in your underwater welding electrodes which we need for different welding jobs.
> Please send us your latest catalogue and price list and some samples of these electrodes. Do you offer any special discounts? Furthermore, we would like to ask what your terms of payment and delivery are. Is there a supplier in Germany or do we have to order the electrodes directly from your office in the USA? We look forward to hearing from you soon.
>
> Yours faithfully
>
> ...
> Schweißtechnik GmbH
> Kiel, Germany

Aufgabe 5 Interaktion   30 VP

Sie befinden sich am Stand einer ...

Folgende Punkte sollten in der Präsentation vorgestellt werden:
Product features
a) Easy portability
b) Four position heat setting
c) Perfect for welding thin metal of aluminium and stainless steel
d) Thermal overload protection (automatic shut-off)
e) Includes: flux core wire, contact tips, ground clamp
f) Wire speed control
g) Runs on standard power
h) External polarity plugs for quick changeover

Workbook · Technical Milestones

9 783128 082691

1 **Put the words into the correct order and put the verbs into the correct tense.**

Example: He / (adjust) / the switch / already – **He has already adjusted the switch.**

1. We / (talk) / to the customer / yesterday

_____

2. They / (repair) / the machine / a week ago

_____

3. I / (order) a multimeter / already

_____

4. We / (send) / your consignment / on June 14ᵗʰ

_____

5. I / (receive) / your mail / just

_____

6. When / you / (send) / the order

_____

7. I / (read) / the manual after he / (suggest) / it

_____

2 **Put the following sentences into correct English.**

1. Wir haben noch keine Bestellung erhalten.

_____

2. Die Probleme mit dem Transformator haben wir gestern bemerkt.

_____

3. Die Anweisungen (instructions), die unser Ausbilder uns vor kurzem gab, waren sehr hilfreich.

_____

4. In der Schulung (training) letzte Woche haben wir viele neue Funktionen des Messgerätes gelernt.

_____

5. Er hat die Diode falsch eingesetzt.

_____

6. Er hat noch nicht bemerkt, dass die Schaltung nicht funktioniert.

_____

C Phrases refresher

1 You want to give instructions to some apprentices about safety rules for electricians. Complete the statements using the expressions from the box.

remember to • be careful with • make sure to • be sure to • it is important to • it is forbidden to • the best thing you can do is to • don't forget to

"Our general rule is: **1** _____ work on energised parts. You all remember the five

safety rules for electricians? Rule number one: de-energise: this means before you work on the mains,

2 _____ turn off the fuse. Rule number two: secure against restarting, meaning

always **3** _____ put a lock on the control cabinet, so that nobody can switch the

fuse back on while you are working. And then our rule number three: test absence of voltage: in other

words, **4** _____ test that the wires are not live. And here comes rule number four:

ensure earthing and short-cutting, but only with voltages above 1000 V. Also for circuits above 1000 V

rule number five applies: **5** _____ other wires next to the one you are working on;

6 _____ cover them with a heavy rubber cloth or fence off with gates. With regards to

informing your colleagues, **7** _____ put up a sign saying "Danger – High voltage." And

after you've finished your job, **8** _____ give a message to your workmates."

2 Put the following sentences into English.

1. Denken Sie daran, Gummihandschuhe anzuziehen.

   _____

2. Es ist wichtig, dass Sie zu jeder Zeit Sicherheitsschuhe tragen.

   _____

3. Seien Sie vorsichtig mit jeder Art von Feuchtigkeit.

   _____

4. Es ist wichtig, dass Sie die Vorschriften genau kennen.

   _____

5. Es ist das Beste, schwierige Reparaturen mit einem erfahrenen Kollegen zu besprechen.

   _____

6. Stellen Sie sicher, dass Sie die Notfallmaßnahmen (*emergency measures*) kennen.

   _____

Module 9 Energy and the environment

A Module refresher

1 Match the sentence beginnings on the left with the endings on the right.

1. Global warming describes	a) storms, melting icecaps and floods.
2. The increase in temperature	b) is mainly caused by emissions from industry and cars, as well as CFCs.
3. The rise in carbon dioxide	c) the average surface temperature of the earth would be around 0°C.
4. The increase in the temperature in a greenhouse	d) can be attributed to the rise in carbon dioxide.
5. Without the greenhouse effect	e) is caused by the glass preventing long-wave radiation from leaving.
6. Some possible effects of global warming are	f) the increase in the average temperature on earth.

2 Match the sentences with the six different forms of power.

A Solar power • B Hydroelectric power • C Tidal power • D Wave power • E Wind power • F Nuclear power

Tiny amounts of waste can endanger all forms of life on earth. ☐

The ebb and flow of the tides can be used to turn a turbine. ☐

Potential gravitational energy can be transformed into other forms of energy. ☐

A single cell produces only very little current. ☐

Huge blades are mounted on top of slim towers. ☐

The water arrives at the turbines at very high pressure because of its great height. ☐

A large area of sea is covered with floats connected together. ☐

It generates huge amounts of electricity from small amounts of fuel (= Brennstoff). ☐

Photovoltaic panels convert light directly into electricity. ☐

Water is stored behind a dam. ☐

It has a reputation for making a swooshing noise day and night.	☐
The nucleus is split roughly in half and releases energy in the form of heat.	☐
The spinning movement turns a generator, which produces electricity.	☐
The kinetic energy caused by up-and-down movements is converted into electricity as in a dynamo.	☐
The difference between low and high tide has to be large enough.	☐
When light shines on silicon layers, an electric current is produced.	☐
A dam is built across the mouth of an estuary.	☐

3   **Read the text on solar power and complete it with suitable words from the box.**

> directly • current • devices • generate • shines • converted • friendly • individual • connected • side

Solar power is an environmentally **1** _____ technology. Sunlight is **2** _____ into electricity without any polluting **3** _____ -effects. We can use the sun in two ways to produce energy: solar cells and solar panels. Photovoltaics (PVs), generally called solar cells, are semiconductor **4** _____ that convert light **5** _____ into electricity. PVs are made of thin slices of silicon. Two different types of silicon layers are **6** _____ by a wire. When light **7** _____ on them, an electric **8** _____ is produced. A single solar cell produces only very little current, approximately 0.5 volts, but when many **9** _____ cells are connected in series, as in a battery, they can **10** _____ a useful amount of power.

4   **Listen to Dr Wall's predictions about future energy consumption and answer the questions.**

A 1.32

1. What does he believe?

   _____

2. What will happen within 10 years?

   _____

3. What will happen within 50 years?

   _____

4. How will we reduce household energy consumption?

   _____

B Grammar refresher

Participles

Das Present Participle *wird mit* Infinitiv + -ing *gebildet.*
Example: Wave power is an **expanding** source of energy.
Das Past Participle *wird mit der* dritten Stammform des Verbs *gebildet.*
Example: After repairing the wind turbines we had to replace three **damaged** blades.
Partizipialkonstruktionen *werden folgendermaßen verwendet:*
– als Adjektive
Example: There is a **broken** generator that doesn't produce electricity anymore.
– zur Verkürzung von adverbialen Nebensätzen mit Konjunktion (when, although, as, while, ...)
Example: While **replacing** the blades, the technician received a phone call.
– zur Verkürzung von adverbialen Nebensätzen ohne Konjunktion
Example: **Dropping** the hammer, he shouted to the others to watch out.
– zur Verkürzung von Relativsätzen
Example: Solar parks **erected** in the desert can supply huge amounts of power.
– zur Verknüpfung von zwei Hauptsätzen
Example: The technician showed us the turbine. He warned us to be careful.
The technician showed us the turbine **warning** us to be careful.
→ **SB p. 236 Participles**

Put the verbs in brackets into the present or past participle.

1. The technician replaced the _____ (break) generator.

2. The train _____ (transport) the waste crashed into the oncoming truck.

3. Peter damaged the transformer _____ (do) the maintenance.

4. The solar panel _____ (find) at the building site belongs to the supplier.

5. The people _____ (develop) the new product are all very experienced.

6. I heard my boss _____ (talk) to the supplier.

7. Scientists always have their results _____ (check) by other experts.

8. We stood there _____ (wait) for the dam to open.

9. He saw his colleague _____ (look) into the turbine.

10. _____ (Have) started the pump, the water flowed into the reservoir.

C Phrases refresher

1 Put in the most suitable words from the box to complete the sentences.

up to • firmly • really can't • up • about • with • due to • on • seems • no doubt • on • feel

1. What do you think _____ the weather forecast?

2. What's your opinion _____ global warming?

3. I'd like to hear your views _____ the greenhouse effect.

4. What I _____ is that it is all being over-exaggerated.

5. It _____ to me that our climate has changed in the last ten years.

6. I _____ believe that something has to be done now.

7. There is _____ that if nothing is done, there will be a major disaster.

8. This is _____ the increase in carbon dioxide levels.

9. I agree completely _____ what you say.

10. I agree _____ a point, but what about the effect on low-lying areas?

11. I'm sorry but I _____ agree with you on that.

12. To sum _____ , there seems to be a major shift in weather patterns, which we cannot predict accurately.

2   **Professor Winter and his team are presenting their research on past and future temperatures on the island of Atlantis over a period of 200 years. Complete his presentation with the help of the graph below and the phrases in the boxes where they fit. Make at least seven statements.**

rise / go up / increase reach a high of remain constant fall / go down / drop / decrease reach a low of	enormous(ly) / dramatic(ally) / sharp(ly) substantial(ly) / considerable / considerably little / slightly hardly

Example: Professor Winter:

"The temperature on Atlantis rose considerably between 1850 and 1860 by 10 degrees Centigrade."

_____

_____

_____

_____

_____

_____

_____

_____

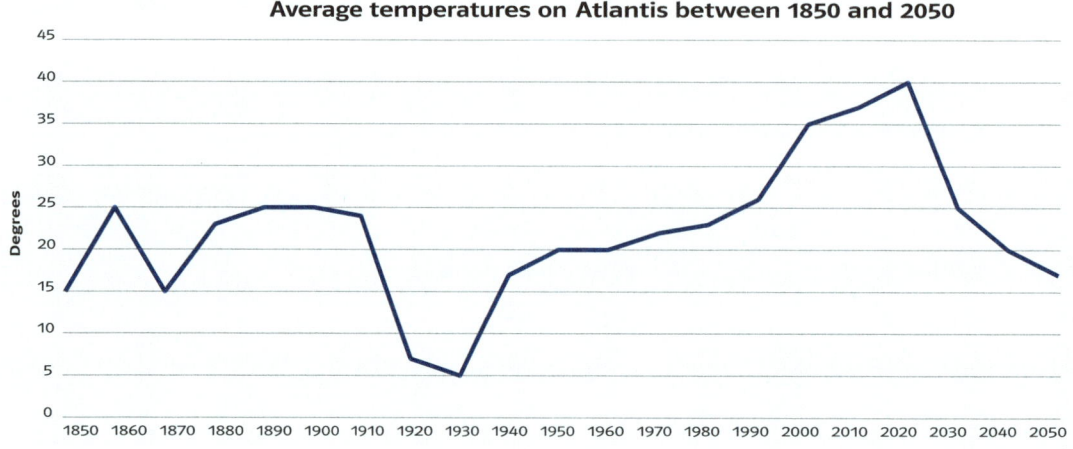

Average temperatures on Atlantis between 1850 and 2050

Module 10   Business trips

A   Module refresher

1   You are preparing for a business trip. Put the activities below in the order they must be carried out.

> **Business trip – to do list**
>
> ☐ Make arrangements for getting to and from the airport in both the city you are leaving from and the one you are visiting.
> ☐ Buy a small guide–/phrasebook about your destination and the language spoken there.
> ☐ Register with ESTA if you are travelling to the USA.
> ☐ Make a list of the appointments you have in the city you are visiting.
> ☐ Call your travel agent to book your flights/book your flights on the internet.
> ☐ Find out what the weather will be like where you are going and plan to pack accordingly.
> ☐ Make sure that your passport/identity card is up to date.
> ☐ Check out what sights or events you might visit if you have free time in the city you are visiting.
> ☐ Make hotel reservations.

2   Listen to the dialogue and complete it.

◎ A 2.2

Consulate: Consulate General of the United States of America, Evans speaking.

Claudia: Claudia Schneider speaking. I'd like some information about **1** _____ and **2** _____ for British and German citizens travelling to the USA this year.

Evans: Well, if you have a machine-readable passport and you plan to stay in the United States for **3** _____ 90 days, and if you have a return ticket, you are eligible for the **4** _____ program, VWP.

Claudia: **5** _____ , what is a machine-readable passport?

Evans: It's one that has two lines at the bottom of the photograph page that can be read by a machine or computer. It must also have a **6** _____ .

Claudia: What about **7** _____ passports, the ones with fingerprints on a chip?

Evans: Biometric passports from VWP countries are **8** _____ since October 26th, 2006. If your passport was issued before October 26th, 2006, and is not biometric but is machine readable, it is still valid for travel.

Claudia: Are there any particular procedures or regulations at the **9** _____?

Evans: The US-VISIT program includes scanning for fingerprints and digital photographs of visitors on entry. This adds a few minutes to the entry procedure. In addition, if you are travelling on the VWP you must apply for **10** _____ to enter the USA by sea or air no later than seventy-two hours prior to your departure. This can be done only online, using

the Electronic System for Travel Authorization (ESTA). There is a fee of fourteen dollars,

11 _____ by credit card. Approval is usually received within minutes and

is valid for two years or until your passport **12** _____ . You must supply

information about yourself, for example name, date of birth, country of birth, passport

information, travel information such as **13** _____ name, flight number, city

where you are **14** _____ , the address where you expect to spend the first

night in the USA, and credit card information. It is suggested that you print out the

15 _____ response and carry it with you.

Claudia: Thank you very much!

3   **Match the sentences or phrases on the left with suitable endings on the right.**

1. We have a reservation	a) when we must check out?
2. Could you tell us	b) for a wake-up call?
3. Welcome to the Excelsior Hotel.	c) a good restaurant?
4. Can you recommend	d) for a double room with shower.
5. The car rental desk	e) May I have your name, please?
6. Shall I arrange	f) is just across the hall.

B   Grammar refresher

If-clauses

Type I, probable (erfüllbar): *if + present + will + infinitive*
Example: If we **are** delayed by traffic, we'**ll miss** our flight.
Type II, improbable (unwahrscheinlich): *if + simple past + would + infinitive*
Example: If you **oiled** this machine regularly, it **would run** for a long time.
Type III, impossible (nicht mehr erfüllbar): *if + past perfect + would + present perfect*
Example: If you **had been** at the meeting, you **wouldn't have missed** the new deadline for your project.
→**SB p. 237 If-clauses**

1   **Complete the sentences below using the correct form of the verbs in brackets.**

1. If the plane is delayed, we (to have) time to go over my sales talk again.

   _____

2. You (to use) much less gasoline if you hadn't driven so fast.

   _____

3. He (not to get) a seat if he comes late.

   _____

4. I won't be surprised if it (to happen) again.

   _____

5. I (not to be) surprised if it happened again.

_____

6. I (to take you out) to a meal if you had more time before your flight.

_____

7. If it (to rain), I'll come home earlier.

_____

8. If they had talked to us earlier, we (to solve) the problem.

_____

9. If you (to be) stronger, you would be able to carry your own suitcase.

_____

2 Put the following sentences into English.

1. Wenn ich diesen Flug nicht gebucht hätte, wäre ich zu spät zur Konferenz gekommen.

_____

2. Wenn wir diesen Vertrag bekommen, geben sie uns einen Bonus!

_____

3. Was wirst du tun, wenn du nicht mit der Geschäftsführerin reden kannst?

_____

4. Das Hotel hätte nicht so viel gekostet, wenn Sie ein Doppelzimmer gebucht hätten.

_____

5. Wenn das Wetter besser wird, werden mehr Menschen die Messe besuchen.

_____

C Phrases refresher

1 What might the passenger (p) and what might the ticket agent (t) say? Fill in a 'p' or a 't' in each box.

1. Is there an afternoon flight to New York on Thursday? ☐
2. Online check-in begins 23 hours before takeoff. ☐
3. I have a window seat and a middle seat available. ☐
4. I'd like to book two seats in economy class, please. ☐
5. I'm afraid the earliest westbound flight departs at 20:33. ☐
6. Yes, there will be a 1-hour stopover in London. ☐
7. Are there any stopovers on this flight? ☐
8. I'd prefer a seat on the aisle. ☐
9. You are allowed a maximum of 20 kg luggage. ☐
10. How much luggage may I take with me? ☐

2 **Complete the following dialogue about booking a hotel room using suitable words or phrases.**

A: I would like to **1** _____

a room with a **2** _____

for three **3** _____ .

We **4** _____ on May 10th

and we **5** _____ on May 13th.

B: Would you like a **6** _____

or a **7** _____ room?

A: A **8** _____ room, please,

with two beds. We don't smoke, so please

make it a **9** _____ room!

B: The **10** _____ is $80 per person, per night, and **11** _____ is included

in the price.

A: How **12** _____ can we check in?

B: You may **13** _____ any time after 3 pm, and **14** _____ is

no later than 11 am.

A: Is there a **15** _____ to the hotel from the **16** _____ ?

B: There is a **17** _____ from the airport to all major hotels in the city centre.

A: Please **18** _____ our booking as soon as possible. May we pay

by **19** _____ ?

3 **Put the following phrases in the right order by giving them numbers.**

☐ Here is the menu.

☐ Would you like to order an hors d'oeuvre?

☐ It was a pleasure to serve you, sir.

☐ Would you like a salad with your entrée?

☐ Will you be having soup?

☐ How about coffee or tea?

☐ How would you like your roast beef done?

☐ Please be seated. I'll bring you water right away.

☐ Can I bring you the dessert menu?

☐ What would you like to drink with your meal?

A Module refresher

1 Read the text below. Then explain the concept of outsourcing / offshoring in German and outline the advantages and disadvantages of the practice.

Outsourcing

One outgrowth of globalisation has been outsourcing, particularly offshore outsourcing or offshoring, a process by which a company contracts one or more of its processes or services to service companies overseas. This may involve one or more stages of an industrial process, software development for a company's needs, or even a call center to handle questions or problems of customers thousands of kilometres away. The company pays them contracted fees for service, instead of employing the people who perform the service and providing them with health insurance and an old-age pension. There are many reasons for outsourcing, among them to avoid regulations on manufacturing and commerce in their own country, to avoid energy costs, and to avoid the cost of labor, which may well be union-regulated in the global North. Outsourcing can also be done within a country or even within a company. In the former case the result has been an explosive growth of temporary employment. In the latter case, a company may simply 'fire' some of its employees and pay them by the hour or on short-term contracts to do exactly what they have been doing, even at the same desk.

While outsourcing was thought to be a positive development around the turn of the 21st century, it has proven to have many negative effects. The global advance of technology has led to more tasks in industry being offshored. While there has been a trend to industrialisation in the Third World, including Central and Eastern Europe, with accelerated urban growth in societies which only a generation ago were chiefly agricultural, the trend in the North (Western Europe and North America) has been one of deindustrialisation. This has meant a shift in the labor market in northern countries from technical jobs to jobs in health care and social services. These cannot be outsourced, but have led rather to large waves of migration from less prosperous countries to fill the need in Western Europe and the USA.

Owing to dissatisfaction with the results of outsourcing in many areas, there is now a countermovement to return many processes and services physically to the parent company. Auto manufacturers have found that some parts have been of such poor quality that they have caused delays in production or dissatisfied customers. And European companies with call centers in the Philippines or India, for example, have discovered that workers who speak the company language with an accent or who come from a different culture are not adequate to help or to pacify irate callers, resulting in loss of business.

2 Find synonyms in the text for the following words and phrases.

1. to get around _____

2. to deal with _____

3. has turned out _____

4. increased _____

5. angry _____

6. poorer _____

3 Listen to the interview. Answer the questions below with full sentences.

A 2.5

1. Why is the interview conducted in English?

_____

_____

_____

2. What position is Mr Rossmann applying for?

_____

_____

_____

3. What did Mr Rossmann study and what did he focus on during that time?

_____

_____

4. What practical experience does Mr Rossmann have?

_____

_____

5. Would Mr Rossmannn have to work in an office?

_____

_____

6. What is the minimum number of hours he would be expected to work per week, and how are they documented?

_____

_____

B Grammar refresher

Future

Will future
Will *drückt die Bereitschaft, etwas zu tun, und Versprechungen aus.*
Example: I'**ll** do my best to get the job.

Future with 'going to'
Going to *wird verwendet, um persönliche Planungen und Vorhaben auszudrücken.*
Example: I am **going to** apply for a job in marketing.

Present continous
Das Present continuous *wird verwendet um eine feste Vereinbarung (gewöhnlich mit Zeitangabe) mitzuteilen.*
Example: Jason **is meeting** his colleague for dinner this evening.

Simple present
Das Simple present *wird für die Zukunft in Bezug auf Fahrpläne, Öffnungszeiten, usw. verwendet.*
Example: The store **opens** at 9 o'clock tomorrow morning.
→ **SB p. 226 Future**

1   **Complete the sentences using the most suitable type of future tense.**

1. The people _____

    (to panic) when they learn who has won the election.

2. The new CEO _____

    (to arrive) from London today.

3. The new department store _____

    (to open) tomorrow.

4. He doesn't agree with our terms so he

    _____ (not to sign) the contract.

5. They _____ (to build) new headquarters this spring.

6. The advertising campaign _____ (to start) on Monday.

7. We _____ (to explore) a new approach to distributing our product.

8. How do you think they _____ (to react) to our offer?

2   **Put the sentences into English using the most suitable form of the future tense.**

1. Sollten wir Kinder bekommen, werde ich aufhören zu arbeiten.

    _____

2. Er wird ab diesem Herbst zu Hause als Übersetzer für den Verlag arbeiten.

    _____

3. Wegen des Streiks wird unser Zug verspätet sein.

    _____

4. Ich denke nicht, dass sie unser Angebot akzeptieren werden.

    _____

5. Sie werden die neue Anzeigen-Kampagne am Dienstag diskutieren.

    _____

6. In der nächsten Sitzung werde ich meine Ansicht verteidigen.

    _____

C Phrases refresher

1   **Join the phrases from the left with the statements on the right. Several combinations are possible.**

1. From my point of view,	a) outsourcing has been bad for the domestic labor market.
2. I am convinced that	b) the European Union is becoming too large.
3. It is a fact that	c) the internet has brought the people of the world closer together.
4. I think everybody agrees that	d) globalisation is more positive than negative.
5. Taking these arguments into consideration,	e) more and more people will choose to work from home.

2 Complete the discussion using the phrases below.

> In my experience • In my opinion • I think you'll find that • I'm afraid I don't agree • This proves that • I see your point, but • Looking at the statistics • As I see it

A: I've read that many large companies are rethinking their approach to eWork.

1 _____ , two-thirds of employers in the USA allow their employees to work from home.

B: **2** _____ the system works, doesn't it?

A: Well, **3** _____ eWorkers are more productive, but they don't come up with as many innovative ideas for their company. **4** _____ , employees need to have personal contact with one another. **5** _____ , the input from discussions over lunch in the cafeteria or in team meetings can generate new approaches to problems.

B: **6** _____ . When people can manage their working time around the needs of their families, they feel less stress and their minds are freer. And they save a lot of time by not having to travel to and from the office.

A: **7** _____ communication face to face with people who are working on the same project is probably more important for its success. **8** _____ , the advantages of having employees working together outweigh the costs saved on office space when employees work at home.

3 Read the advertisement and write a letter to apply for the job offered.

ww#.softapps….co.uk

We are looking for a

Software Application Writer

to join our small staff.

You should have some experience with document management software and business workflow software. The job entails some travel to clients in England and Germany and you should therefore be fluent in both languages. Salary is negotiable, medical benefits are included.

Submit written application to:
SoftApps Ltd.
154 Ridgley Terrace
London W 14
email: carter@softapps….co.uk

A Module refresher

1 Sitcomfy presents itself in the company brochure below. Complete the text.

> demanding • fold-flat • mission • adjustable • required • exceed • locations • seating preferences • storage • ventilation • fold-and-tumble seating • develop

At over 60 **1** _____ around the world, our 26,000 employees, including over

2,000 engineers and designers, **2** _____ and produce not only the seating systems,

but also much of the automotive electronics and electronic energy management systems

3 _____ . Seats nowadays are very complex products. With people spending increasing

amounts of time in their vehicles, comfort is not just a matter of opinion. Because every person has

their own individual **4** _____ ,we have had to develop highly flexible, truly

5 _____ seating systems especially for the **6** _____ premium segment.

A single global process rapidly transfers best practices throughout our organisation. So our customers

get best-in-class products no matter where in the world we produce. We share a single

7 _____ , and that is to **8** _____ our customer expectations at all times.

I certainly hope we managed to do this last year with every one of our 7.5 million installed seating

systems worldwide. We cover the full spectrum –
from simple two-way seating to twenty-way seating
systems for luxury vehicles. These specialised
products include **9** _____ systems,
massage functions, and humidity-controlled seating.
We also specialise in rear **10** _____ and
11 _____ seating that offers solutions
for more space and **12** _____ .

2 Listen to the dialogue and decide whether the following statements are True or False.

A 2.7

	True	False
1. Manfred is calling from Porto.	☐	☐
2. The caller is phoning about alterations that were made to a specific seat.	☐	☐
3. The new seat assembly process has always been having problems.	☐	☐
4. Grace Cooke tested the right housing.	☐	☐
5. CoolCo increased the width after agreement with the design department.	☐	☐
6. The reinforced bracket is obstructing the new housing.	☐	☐
7. Manfred should ring Mike in Porto first.	☐	☐

3 Underline the correct word or phrase of the words in bold.

To:	Manfred.Schuster@shark.co … .com
From:	Grace.Cooke@shark.co … .com
Copy:	Mike.Woodberry@shark.co … .com
Date:	15/06/20___
Subject:	Packaging modification issues for product no. B-172

Dear Manfred,

here is the progress **1** **report / writing** you were asking for. The **2** **changements / modifications** on the retaining bracket for the cooling fan are going **3** **referring to / according to** schedule. The bracket has been reinforced to **4** **simplify with / comply with** crash-test specifications and will now **5** **conform / confirm** to the new European standard. Mike Woodberry has already received our newly released samples and is **6** **producting / manufacturing** the first batches which will be sent to you by 30ᵗʰ June. As agreed, the bracket is to be **7** **mounted / disassembled** onto the centre crossbar under the main seat pan. We have **8** **experimented / researched** with the new bonding method and we haven't found any **9** **flaws / wrongs** yet. Nevertheless, I am not sure if it will have a negative impact on the assembly process. Mike – will you let us know about that? And please recheck the dimensions for the packaging. The digital **10** **lock-ups / mock-ups** in our system looked fine and I don't expect any problems, but there is always a chance of an unforeseen **11** **setback / backout** and our milestone plan is very tight. As you know, the **12** **deadline / lifeline** for delivery to the customer is 20ᵗʰ July.

Regards,

Grace

4 Complete the production process of a chassis using the words in the box.

> marriage • chassis • deep-drawn • fluids • body-shells • immersion • conveyed • sheet metal • dipped • sub-assembly • tack-welded • booths

1. Here the _____ is cut and _____ to form the outside and inside body panels.

2. The parts are then _____ in a framing station and robots weld the parts together.

3. The _____ are _____ to the area where they are _____ into a huge _____ bath, sprayed in spray _____ and dried in ovens.

4. The next process manages the underbody _____ and engine build-up.

5. The _____ of the body and _____ takes place on the final line.

6. Here all of the _____ are filled, e.g. for the power steering and for the radiator; inspection and quality testing is also carried out.

B Grammar refresher → WB p. 19 Questions and SB p. 228 Questions

Ask questions about the phrases in blue using the question words given.

1. Manfred calls from Porto every Friday. (Ask: Where?)

   _____

2. We are meeting John at 3 o'clock. (Ask: What time?)

   _____

3. He gave the report to Helen. (Ask: Who?)

   _____

4. Mike didn't install the bracket. (Ask: Why?)

   _____

5. They have invested one million up to now. (Ask: How much?)

   _____

6. CoolCo produced the units. (Ask: Who?)

   _____

7. Grace sent the samples when she received the confirmation. (Ask: What … when … ?)

   _____

8. The product isn't going to make a profit. (Ask: Why?)

   _____

9. They are going to need 3 hours for the job. (Ask: How long?)

   _____

10. We would redesign the product if we had more time.

    (Ask: What … do … ?)

    _____

C Phrases refresher

1 **Match the sentence beginnings on the left with the endings on the right.**

1. Would you like me to check	a) reinforcing the bracket?
2. I'd like you to recheck	b) me the results by tomorrow?
3. How about mounting	c) the screws for you.
4. Do you think you could send	d) the bracket onto the centre crossbar?
5. Is it all right if I inform	e) me what went wrong?
6. Let me tighten	f) them about the impact later?
7. Would you mind telling	g) the packaging as soon as possible.
8. Why don't you try	h) the specifications for you again?

2   **Match the phrases with the four categories.**

a) We can't all speak at the same time.
b) One at a time, please.
c) Let's get started.
d) Do we all agree on that point?
e) Over to you, Tom.
f) See you all at the next meeting on June 1st.
g) Thank you for taking part.
h) I'm glad you are all able to attend.

i) The aim of the meeting is to decide on the product strategy.
j) The first point on the list is the market analysis.
k) Could we stick to the point, please?
l) Right, now let's move on to the next point.
m) Mike, it's your turn.
n) John is excused because he is on holiday.
o) Good morning everybody.
p) I didn't catch what you said.
q) Would you mind repeating it, please?

Beginning a meeting	Outlining the agenda	Running a meeting	Finishing a meeting

3   **Put the phrases / sentences into German.**

1. The company is divided into different departments.

_____

_____

2. The head of a company is called managing director, general manager or chief executive officer (CEO).

_____

_____

3. Each department is split into teams which are led by team or group leaders.

_____

_____

4. The head of a department is called department manager.

_____

_____

Module 13 Control technology

A Module refresher

1 Match the devices with the definitions below.

solenoid • lobe pump • pressure switch • valve • proximity sensor • switch • limit switch • actuator

1. _____: This is a device used to detect the presence of nearby objects without any physical contact. It often emits an electromagnetic or electrostatic field, or a beam of electromagnetic radiation and looks for changes in the field or return signal.

2. _____: This is a device that supplies and transmits a measured amount of energy for the operation of another mechanism or system. It is frequently used as a mechanism to introduce motion, or to clamp an object to prevent motion.

3. _____: This is a device that automatically cuts off current to an electric motor when an object moved by it, such as an elevator or a production line, has passed a given point.

4. _____: This is a device for turning on or off or directing an electric current or for making or breaking a circuit. It also refers to a small lever or button on computer hardware.

5. _____: This is a device used to move gases, liquids or slurries. Fluid flows around the interior of the casing. Shaft support bearings are located in the gearbox, and since the bearings are out of the pumped liquid, pressure is limited by the bearing location and the shaft deflection.

6. _____: This is a device for stopping or controlling the flow of a liquid, gas, or other material through a passage, pipe, inlet, outlet, etc.

7. _____: This is a device that makes electrical contact when a certain set pressure has been reached on its input. It is used to provide on/off switching from a pneumatic source. It monitors a process's pressure by applying the pressure to a piston or diaphragm (the sensing element), which generates a force.

8. _____: This is device that acts as an electric conductor wound as a helix, or as two or more coaxial helices, so that current through the conductor establishes a magnetic field within the conductor. It is also used as a switch in which a metal rod moves when the current is turned on. It is often used in automotive starting systems.

2 Frank Walker explains to his apprentices how relays work. Complete his presentation.

voltage is applied • logical control • when the input coil • resulting current • without a mechanical switch • when the coil is energised • pulls a metal switch • magnetic field • contacts touch • relays

"A relay is a simple device that uses a **1** _____ to control a switch. Relays allow

power to be switched on and off **2** _____ . When **3** _____ to

the input coil, the **4** _____ creates a magnetic field. The magnetic field

5 _____ towards it and the **6** _____ , closing the switch.

The contact that closes **7** _____ , is called 'normally open'. The 'normally closed'

contacts touch **8** _____ is not energised. It's quite common to use

9 _____ to make simple **10** _____ decisions."

3 Listen to a recommendation of a new robot. Decide whether the following statements are True or False.

A 2.10

	True	False
1. The robot is made entirely of carbon fibre.	☐	☐
2. Heavier materials have a negative impact on the customer's throughput.	☐	☐
3. The robot has a flexibility rate similar to cam-driven mechanisms.	☐	☐
4. Its construction makes high speeds in large areas possible.	☐	☐
5. All the motors are in the base.	☐	☐
6. The most of the robot's weight is in the arm.	☐	☐
7. Because of its work envelope it can support different palletizing operations.	☐	☐
8. The robot can carry up to 40 kg.	☐	☐

4 Put the phrases on negotiating strategy in the right order by giving them numbers.

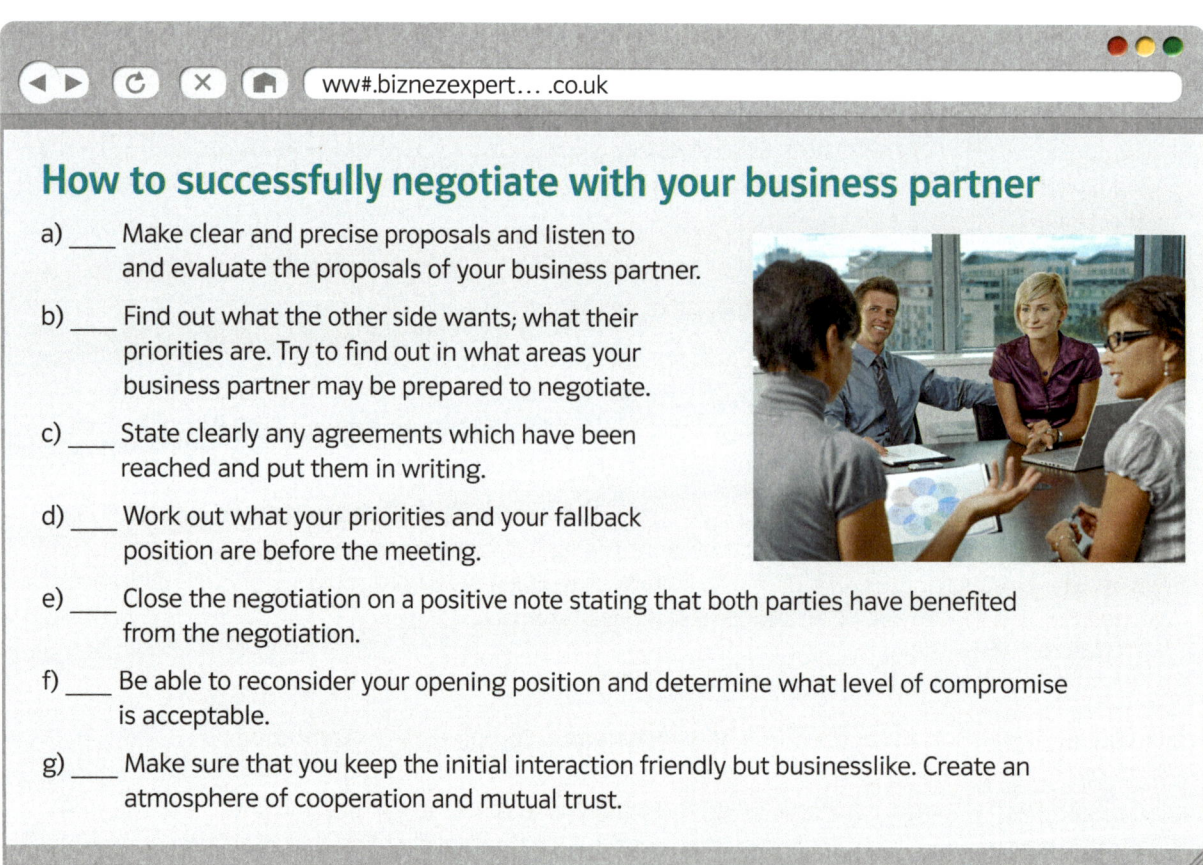

ww#.biznezexpert… .co.uk

How to successfully negotiate with your business partner

a) ____ Make clear and precise proposals and listen to and evaluate the proposals of your business partner.

b) ____ Find out what the other side wants; what their priorities are. Try to find out in what areas your business partner may be prepared to negotiate.

c) ____ State clearly any agreements which have been reached and put them in writing.

d) ____ Work out what your priorities and your fallback position are before the meeting.

e) ____ Close the negotiation on a positive note stating that both parties have benefited from the negotiation.

f) ____ Be able to reconsider your opening position and determine what level of compromise is acceptable.

g) ____ Make sure that you keep the initial interaction friendly but businesslike. Create an atmosphere of cooperation and mutual trust.

5 Listen to the dialogue and complete it.

A 2.11

Mark: I believe that D2R2 is quite new on the market.

Peter: Yes, that's right but our **1** _____ have been in the business for over 20 years.

Mark: And how's business doing?

Peter: Well, actually, it's growing nicely. We have received a few **2** _____ and, to be honest, that's all we are looking for at the moment. Your order fits in perfectly with our present strategy. While we are expanding capacity, we don't want to **3** _____ quality. As you may know, we have just been given the top-quality award for specialized applications.

Mark: Yes, we were **4** _____ by that. And your prices are **5** _____ as well. Where we do see some difficulty is with the delivery period. Your standard delivery time is 90 days.

Peter: Oh, I thought that wasn't **6** _____ , since it's standard in our industry. I'm not sure but I **7** _____ to deliver 80–90 days after confirmation of our offer. But I'd have to get back to you to confirm that.

Mark: You mean it could take the full 90 days?

Peter: No, no, not at all. **8** _____ reduce the time.

Mark: Hmm. We were looking at 60 days, actually. We are tooling up for a large order for the US market and if we don't **9** _____ , we will certainly have to pay a penalty.

Peter: 60 days is very tight and you know that 90 is standard in the industry but if we shifted other orders and **10** _____ a few extra technicians, we should be able to manage it. I'll have to get back to you tomorrow on that. I need to talk to our production manager to see if it's possible. **11** _____ compensate us for any extra costs involved?

Mark: Hmm, possibly, but **12** _____ delivery guarantees. **13** _____ order a second robot with special applications, for delivery in 180 days, same price? Would that be compensation enough?

Peter: Now, that's an interesting proposal. I think we will definitely be able to do business. We'll have to **14** _____ but, as there are likely to be **15** _____ for a long-term relationship, it will be worth it.

B Grammar refresher

Possibility, probability and certainty

Im Englischen lassen sich verschiedene Abstufungen von Wahrscheinlichkeit *folgendermaßen ausdrücken:*

Examples:

100 % = certain, must be
75 % = probable, may well be, likely
50 % = possible, may be
25 % = improbable, highly unlikely
0 % = impossible, can't be

→ **SB p. 189 Possibility, probability and certainty**

Choose the correct terms related to possibility, probability and certainty and put them in the gaps below.

likely to • will probably • can't possibly • possible • certainly won't • may well • highly unlikely • may have been • impossible • must have been

1. It was _____ to meet the production targets because of the strike.

2. The actuator _____ have failed. It was brand new.

3. It is _____ that the switch will wear out after five years.

4. The machine _____ break down if it isn't kept oiled.

5. The engine _____ repaired before production could start yesterday.

6. It is _____ that we will produce more next year. There's no demand.

7. There _____ a defect in the pump. We don't really know.

8. We are _____ carry out repairs again on the vehicle in the future.

9. The price _____ rise if there is a shortage.

10. The company _____ invest more money in the project. It doesn't have a future.

C Phrases refresher

Compare the valves using the phrases below. Make five sentences. Several solutions are possible.

… larger than …/… the largest • … considerably heavier than …/… the heaviest • … probably less/ more expensive than …/… the most expensive … • usually less/more effective than …/… the least/ most effective (for a specific purpose) … • … better/best because …

Example: **The butterfly valve is probably less expensive than the gate valve.**

ball valve

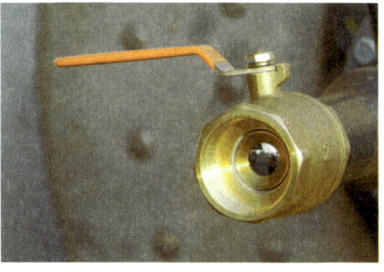

gate valve

butterfly valve

_____

_____

_____

_____

_____

_____

_____

Das KMK-Fremdsprachenzertifikat

Dieser Teil des Workbooks bietet einzelne, nicht zu schwere Aufgaben zu den KMK-Kompetenzen an, die die entsprechenden Aufgaben im Schülerbuch ergänzen und weiteres Übungsmaterial bieten. Bei den Aufgabensätzen steht der Übungscharakter im Vordergrund. So orientieren sie sich an den Vorgaben für KMK-Prüfungen, stellen aber keine Originalprüfungen dar. Eine umfassende Vorbereitung auf die KMK-Prüfung ist hiermit gewährleistet.

Nach dieser Einführung in die Kompetenzen, die das KMK-Fremdsprachenzertifikat überprüft, folgt ein Übungsteil mit Einzelaufgaben für jeden Aufgabentyp.

Der Gemeinsame Europäische Referenzrahmen (GER)

Grundlage der KMK-Zertifikatsprüfungen ist der *Gemeinsame Europäische Referenzrahmen für Sprachen: Lernen, Lehren und Beurteilen, GER (Common European Framework of Reference for Languages: Learning, Teaching, Assessment,* abgekürzt CEF). Der Europarat hat darin unter anderem die vier Kompetenzbereiche **Rezeption, Mediation, Produktion** und **Interaktion** vorgestellt, nach denen in Europa Fremdsprachenerwerb beschrieben werden kann. Außerdem hat er Kriterien entwickelt, nach denen sich der Fortschritt des Spracherwerbs messen lässt: Der Referenzrahmen unterscheidet drei Bereiche: Die elementare Sprachverwendung (A), die selbstständige Sprachverwendung (B) und die kompetente Sprachverwendung (C). Innerhalb dieser drei Bereiche wird nochmals in jeweils zwei Stufen unterschieden, d.h. in A1 und A2, B1 und B2, und C1 und C2, d.h. insgesamt sechs Niveaustufen. Diese Niveaustufen beschreiben, wie einfach oder schwierig die jeweiligen fremdsprachlichen Sachverhalte und Gesprächssituationen sind, die ein Prüfling auf einer bestimmten Niveaustufe bewältigen muss.

Die Umsetzung des GER in den Niveaustufen des KMK-Fremdsprachenzertifikats

Die Kultusministerkonferenz hat aus diesen sechs Niveaustufen zunächst drei ausgewählt, nach denen Englischkenntnisse in Deutschland an beruflichen Schulen zertifiziert werden können, nämlich die Stufen A2 (Stufe I), B1 (Stufe II) und B2 (Stufe III). Die später eingeführten Niveaustufen C1 und C2 werden im gewerblich-technischen Bereich je nach Bundesland selten angeboten. Die KMK-Fremdsprachenzertifikatsprüfung gibt es nur an beruflichen Schulen und wird zur freiwilligen Teilnahme angeboten.

Die vier Kompetenzbereiche des KMK-Zertifikats

Rezeption Für diesen Kompetenzbereich muss ein berufsbezogener englischsprachiger Sachverhalt auf seinen Informationsgehalt hin ausgewertet werden, d.h. man muss wiedergeben können, was beispielsweise in einem Text steht oder in einem Gespräch zu hören ist, welche Informationen für einen selbst oder den Arbeitgeber wichtig sind. In der KMK-Prüfung wird die Kompetenz Rezeption immer durch eine Hörverstehens- und eine Leseverstehensaufgabe überprüft.

Die Hörverstehensaufgabe steht in der Regel am Anfang der Prüfung. Der englischsprachige Text wird zweimal von einer CD vorgespielt. In den vorliegenden Übungen hat der Übende natürlich den Vorteil, die Aufnahme auch öfter anhören zu können. Je nach Niveaustufe sind diese Sachverhalte einfacher oder komplizierter. Die gesprochenen Mitteilungen werden eher langsamer oder auch schneller von einem *native speaker* gesprochen. Hierbei handelt es sich um Mitteilungen auf Anrufbeantwortern, um Dialoge oder Fachvorträge. Die Hörverständnisaufgabe kann zweigeteilt sein, z.B. in eine Aufgabe zu einem Anrufbeantworter und einer Auswertung eines Fachvortrages oder Dialoges. Die hier vorgesehene Punktevergabe kann von den Originalaufgaben abweichen. Bitte beachten Sie, dass in den Original-KMK-Prüfungen z.T. zwei Hörverstehensaufgaben vorkommen können.

Beim Leseverstehen gibt es unterschiedliche Textarten und Aufgabentypen. Der Inhalt eines englischen Fachtextes (Betriebs- oder Montageanleitungen, Garantieerklärungen, Produktbeschreibungen usw.) wird mit deutschen oder englischen Fragen zum Text ausgewertet. Die Beantwortung der Fragen kann auch in speziellen Tabellenblättern, Formularen oder Lückentexten erfolgen. Eher selten wird der Inhalt eines Textes mit *true/false questions* abgefragt.

Mediation Die Mediation verlangt die Übertragung eines Sachverhaltes in die jeweils andere Sprache. Eine wörtliche Übersetzung ist nicht gewünscht, sondern vielmehr eine sprachlich angemessene Übertragung. Wichtig allein ist, ob eine Auszubildende/ein Auszubildender eine Nachricht, eine

Anleitung, ein Handbuch oder eine Verfahrensanweisung so in die andere Sprache übertragen kann, dass beispielsweise Mitarbeiter, die kein Englisch sprechen, die Nachricht umsetzen oder nach der Anweisung verfahren können. Der umgekehrte Fall gilt auch: Englischsprachige Kunden oder Geschäftspartner sollten beispielsweise eine vom Deutschen ins Englische übertragene Verfahrensanweisung richtig verstehen und umsetzen können. Die Übertragung in die andere Sprache muss inhaltlich richtig, aber auch sprachlich so gut nachvollziehbar und unmissverständlich sein, dass sie ohne umfangreiche Bearbeitung oder Hilfe verwendet werden kann.

Produktion Beim Aufgabentyp Produktion geht es um das Verfassen eines Schriftstücks in der Fremdsprache. Je nach Niveaustufe (d. h. Schwierigkeitsgrad) gehört zu diesem Aufgabentyp die Erstellung berufstypischer E-Mails, Faxe, Anfragen, einfacherer und komplexerer Geschäftsbriefe, das Verfassen von Montage- oder Installationsanleitungen für Kunden, das Erklären von graphischen Darstellungen, Verfahrensanweisungen etc.

Interaktion Die Interaktion, also die Kompetenz, die Gegenstand der mündlichen Prüfung ist, legt ihren Maßstab daran an, ob der Prüfling „berufsrelevante Gesprächssituationen bewältigen" kann. Dies reicht von der Erfassung berufsrelevanter Sachverhalte bis zu Situationen, in denen die Gesprächsinitiative ergriffen wird, wo der Prüfling Sachverhalte ausführlich erläutert und Standpunkte darlegt und verteidigt. Die mündliche Prüfung kann als Einzel- oder Gruppenprüfung durchgeführt werden.

Die KMK-Prüfung

Die Fremdsprachenzertifikatsprüfung hat einen schriftlichen und einen obligatorischen mündlichen Teil. Die Punkteverteilung sieht in der Regel folgendermaßen aus:

Schriftliche Prüfung	Mündliche Prüfung
Rezeption 40 VP *	Interaktion 30 VP
Mediation 30 VP	
Produktion 30 VP	
Summe 100 VP	

* Davon entfallen in der Regel 20 VP auf das Hörverstehen und 20 VP auf das Textverständnis. Die Punktzahl für die Hörverstehensaufgabe wird evtl. unterteilt in 8 oder 10 VP für eine Aufgabe zu einem Anrufbeantworter und 12 oder 10 Punkte für die Auswertung eines Dialogs oder Vortrages.

Bei der schriftlichen Prüfung ist eine leicht veränderte Gewichtung der Prüfungsteile innerhalb bestimmter Grenzen möglich.

Die mündliche Prüfung kann als Einzel- oder Gruppenprüfung durchgeführt werden.

Zum möglichst situationsgerechten Rahmen dieser Prüfung gehört natürlich auch, dass für die Prüfung ein **zweisprachiges allgemeinsprachliches Wörterbuch** (kein Fachwörterbuch) zur Verfügung steht, so wie im Betrieb u. U. auch. Jedoch sollte bedacht werden, dass das Nachschlagen von Vokabular wertvolle Prüfungszeit kosten kann.

Die **Prüfungszeiten** sind für die jeweiligen Niveaustufen unterschiedlich geregelt.

Niveaustufe	Schriftliche Prüfung	Mündliche Prüfung
I (A2)	60 Min.	10 Min. pro Prüfling
II (B1)	90 Min.	15 Min. pro Prüfling
III (B2)	120 Min.	20 Min. pro Prüfling

Zum **Bestehen der KMK-Zertifikatsprüfung** müssen beide Prüfungsteile, der mündliche wie der schriftliche, separat bestanden werden, d. h. in beiden Prüfungen muss jeweils mindestens die Hälfte der Punkte erreicht werden. Ein Ausgleich zwischen beiden Prüfungsteilen ist nicht möglich, man kann also eine schlechte mündliche Leistung nicht durch eine gute schriftliche ausgleichen oder umgekehrt. Auch gibt es für die Prüfung keine Anmeldenoten. Im eigentlichen Zertifikat stehen nur die erreichten Punktzahlen, keine Noten.

| **Aufgabe 1** | Rezeption – Hörverstehen | **20 VP**

Hören Sie sich den Dialog zweimal an und ergänzen Sie die folgenden Sätze.

1. The headstock is started by switching on the _____ .   VP 02

2. The power is transmitted to the different _____.   VP 02

3. The _____ rotates the workpiece.   VP 02

4. The _____ drives _____, which rotates the workpiece.   VP 04

5. The workpiece can be rotated clockwise or _____ .   VP 02

6. The _____ speed is set with the speed selection lever.   VP 02

7. The feed of the cutting tool or workpiece is controlled by the _____.   VP 02

8. The _____feeds the cutting tool.   VP 02

9. The workpiece is clamped in the _____.   VP 02

| **Aufgabe 2** | Rezeption – Leseverstehen | **20 VP** |

Lesen Sie den folgenden Text und beantworten Sie die Fragen auf Deutsch.

Apprentices learn how to carry out easy cutting operations with hand tools and power tools in a general workshop. For example, they cut metal bars to certain rough dimensions. Then, they use different types of files to finish the bars to the required dimensions.

Two apprentices are doing their training at a machine tool company in Manchester. After introducing the week's training plan, their instructor asks them to make a square base plate according to the technical drawing in their work plan.

Before starting the job, they clamp the stock in a bench vice so that it is held tightly.

First they cut a metal plate to a certain length and width given in the technical drawing of their operation plan. This cutting job can be done with a hacksaw, an angle grinder or even a circular saw, but very often the trainees prefer to do this job with the grinder because it is easier compared with the physical strength required when using a hacksaw. But while using any of these tools, it is important to wear a pair of goggles to protect eyes from the chips that fly around during the cutting process.

After that the trainees use different files to work the metal plate to a precise length and width. The plate has a precision of a tenth of a millimetre and four right angles. The edges of the plate are deburred with a file. These jobs are started with a roughing file and continued with a finishing file. Now the apprentices check the right angles of the four corners with a square. Then they use a scriber to mark four holes with a given diameter in each corner of the metal plate. After centre punching the marks with a centre punch and a hammer, the trainees are ready to bore the holes with a twist drill.   (309 words)

1. What basic job has to be done before starting the cutting jobs on the material?   4 VP
2. How can the metal plate be cut to the correct length and width?   3 VP
3. What is the advantage of using an angle grinder compared to a hacksaw?   2 VP
4. What is the idea of using a square?   2 VP
5. How are the edges of the plate worked and which tools are used for that job?   3 VP
6. What tools are used to bore the holes in the corners of the base plate?   2 VP
7. What is done before drilling the holes in the corners of the base plate and which tools are used? 4 VP

Aufgabe 3	Mediation	30 VP

Die Auszubildenden der englischen Niederlassung Ihrer Firma besuchen Ihre Werkstatt. Sie stellen dem Besuch die nötigen Sicherheitsvorkehrungen in der Grundwerkstatt, der Dreherei und der Schweißerei vor. Erklären Sie auf Englisch die Funktion der folgenden deutschen Schutz- und Sicherheitsausrüstung.

„Beim Betreten der Werkstätte in unserer Fabrikhalle tragen wir immer Arbeitskleidung, z.B. einen Overall, und Arbeitsschuhe als Schutz gegen herunterfallende Metallteile.
In der Grundwerkstatt werden wir für Arbeiten am Schraubstock ausgebildet, z.B. für Feil- oder Schleifarbeiten an Metallen. Dabei ist es wichtig, unsere Augen mit einer Schutzbrille zu schützen. Bei Bohrarbeiten an der Ständerbohrmaschine trage ich immer ein Haarnetz wegen meiner langen Haare.
In der Schweißerei tragen wir Arbeitshandschuhe zum Schutz gegen scharfe Metallkanten und gegen Hitze beim Schweißen. Wir schützen auch unsere Augen mit Schutzmaske oder Schweißer Handschild gegen das gleißende Licht und den Funkenflug."

Aufgabe 4	Produktion	30 VP

Zusammen mit Ihrem englischen Kollegen bekommen Sie die Aufgabe, auf Englisch eine kurze Handreichung der Arbeitsprozesse zu erstellen, die nötig sind, um den Maschinenschraubstock in der Zeichnung herzustellen. Berücksichtigen Sie dabei detailliert die einzelnen Bauteile des Schraubstocks.

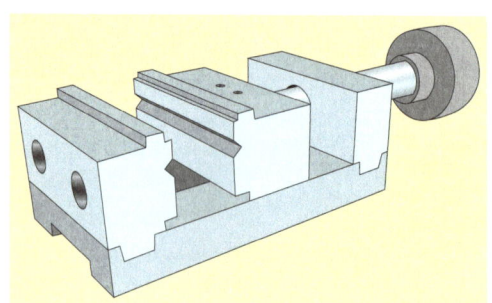

Aufgabe 5	Interaktion	30 VP

Sie arbeiten stundenweise in einem Elektrogeschäft. Ein englischer Kunde interessiert sich für eine Schlagbohrmaschine. Geben Sie das Verkaufsgespräch auf Englisch wieder. Sprecher A beginnt.

Sprecher A	Sprecher B
Begrüßen Sie den Kunden und fragen Sie, ob Sie ihm weiterhelfen können.	Sagen Sie, dass Sie eine Bohrmaschine suchen.
Fragen Sie, wozu der Kunde die Maschine braucht. *(What ... for?)*	Sagen Sie, dass Sie einige Arbeiten in Ihrer neuen Wohnung machen möchten.
Empfehlen Sie einen Akku-Schlagbohrer. Zeigen Sie Ihr neuestes Modell.	Fragen Sie nach den Besonderheiten des Gerätes.
Nennen Sie den kraftvollen Motor mit einem 18 V-LI Akku und einem kompakten Umlenkgetriebe.	Fragen Sie, ob das Modell auch einen Rechts-Links-Lauf hat.
Versichern Sie diese Eigenschaft. Fügen Sie hinzu, dass das Gerät über einen pneumatischen Schlagmechanismus verfügt.	Erkundigen Sie sich nach dem Werkzeugwechsel, z.B. ob er schnell erfolgt. *(What about ...?)*
Antworten Sie, dass das neue Modell über eine SDS-plus Werkzeugaufnahme verfügt.	Fragen Sie, ob Sie auch Löcher mit einem Durchmesser von 10 mm bohren können.
Bestätigen Sie, dass die Maschine Löcher mit einem Durchmesser bis zu 12 mm bohrt bei optimaler Lastdrehzahl.	Ergänzen Sie, dass Sie das schon von dem Vorgängermodell kennen *(previous model)*. Fragen Sie nach weiteren Modellen.

| **Aufgabe 1** | Rezeption – Hörverstehen | **15 VP**

Bernd und Oli, zwei deutsche Auszubildende, und Amélie, eine Praktikantin aus Frankreich, sprechen mit ihrem Ausbilder über verschiedene Schweißausrüstungen. Hören Sie sich das Gespräch zweimal an und beantworten Sie die Fragen auf Deutsch in ganzen Sätzen.

1. What is the main difference between oxyacetylene welding and arc welding? 　　　　　　　2 VP

_____

2. How is the gas mixture generated for oxyacetylene welding? 　　　　　　　2 VP

_____

3. Describe the different parts of gas welding equipment and the functions of these parts. 　　7 VP

_____

_____

_____

_____

4. Which are the main parts used in arc welding? 　　　　　　　4 VP

_____

| **Aufgabe 2** | Rezeption – Leseverstehen | **15 VP** |

Ihre englische Partnerfirma Turntec Ltd. hat für die Benutzer ihrer Drehmaschinen Sicherheitshinweise entworfen, deren Einhaltung bei Dreharbeiten empfohlen werden. Lesen Sie den folgenden Text durch und beantworten Sie anschließend die Fragen auf Deutsch.

Lathe safety
You are responsible for your own safety and the correct operation of your machine. A lathe, like any power tool, can be dangerous if not operated correctly. If metal working is new to you, it is useful to follow good safety practices as given below:

- Always wear eye protection, either a pair of safety glasses with side-shields or a pair of goggles. During the lathe operation, sharp, hot metal chips can fly off at speed or metal spirals can spin off. So be very careful to protect your eyes.
- It is recommended to wear short-sleeve shirts or shirts with tight long sleeves. Loose sleeves can be hazardous if they get caught in the machine while doing rotating work and your hand or arm can be pulled in.
- The same goes for long hair. Wear a hair net or tie back long hair to prevent it from being caught in the rotating tool. Otherwise you may receive a severe injury to your head.
- Wear leather safety shoes to protect your feet from falling heavy material or tools and from sharp metal chips on the shop floor.
- Don't wear jewellery such as chains, necklace, wrist watches or finger rings because they can catch on rotating work and badly injure your hand and fingers.
- Before starting the lathe, always make sure your work piece is properly clamped in the chuck. It is recommended to start the working spindle at low speed and increase it gradually.
- Always remove the chuck key immediately after use. The chuck key can turn into a deadly bullet if the lathe is started with the chuck key in the chuck.
- Always keep your hands away from the rotating work and cutting tools. Some machine operators risk bad cuts when they try to quickly remove metal spirals as they form at the cutting tool.
- When controlling whether the coolant is flowing properly during the cutting operation, avoid reaching across the spinning chuck. You risk touching with the rotating chuck or having the sleeve of your working clothes getting caught in the spinning chuck. (354 words)

1. What special features must safety glasses have? (2VP)

_____

_____

2. Why must safety glasses have these features? (2VP)

_____

_____

3. Why are long sleeves and long hair dangerous? (2 VP)

_____

_____

4. Which properties should safety shoes have and why? (2 VP)

_____

_____

5. Why is it recommended not to wear jewellery? (2 VP)

_____

_____

6. What has to be considered before starting work? (2 VP)

_____

_____

7. How can you avoid the chuck key turning into a deadly bullet? (1 VP)

_____

_____

8. Which additional safety aspect is important when working near a rotating chuck? (2 VP)

_____

_____

Aufgabe 3	Mediation	30 VP

Ihre Firma hat ein Fax von Mike Masters aus England erhalten. Sie werden gebeten, den Inhalt in den wichtigsten Punkten auf Deutsch in ganzen Sätzen schriftlich wiederzugeben.

From: Mike Masters, Machine Tool Centre (MTC), Leeds   To: Hans Mustermann   Fax: +49 711 546 154 12   Date:

Dear Hans

You asked for a few tips on how to test the hardness and tensility of steel components. I hope that the following ideas will help you.

First, check what kind of application the material is needed for because in mechanical engineering it is important that the material does not fracture during the machining process, for example.

Thus it is important to determine the toughness of a metal or in other words, its ability to resist a sudden impact. An impact test, which uses a hammerhead, tests the toughness of a material and measures the energy required to break it.

If you want to measure the tensile strength of a material, you can do a tensile test. This test shows what happens when a material, such as steel, is loaded in tension to the point where it breaks.

The most frequently applied test in industry, however, is the Rockwell hardness test. Hardness is defined as the resistance to indentation and this test measures the permanent depth of an indentation made by an indenter under a certain amount of pressure.

Of course, all of the test applications that I have described only involve samples which have the same properties as the components used in real machining operations.

I hope that this is enough information to get you started. Let me know if you have any more questions.

Kind regards

Mike

Aufgabe 4	Produktion	40 VP

Für einen Auftrag möchten Sie bei der Firma SubWeld entsprechende Elektroden bestellen. Verfassen Sie eine förmliche E-Mail, die die folgenden Punkte enthält.

– Schreiben Sie der Firma, dass Sie Elektroden für verschiedene Unterwasserschweißarbeiten benötigen.
– Bitten Sie um Zusendung des neuesten Katalogs mit Preisliste.
– Erfragen Sie auch, welche Sonderrabatte die Firma gewährt, und wo Sie die Elektroden beziehen können, ob es einen Lieferanten in Deutschland gibt oder ob Sie direkt am Firmensitz in den USA bestellen sollen.

Aufgabe 5	Interaktion	30 VP

Sie befinden sich am Stand einer Industriemesse in Manchester und repräsentieren Ihre Firma ST (Schweißtechnik) GmbH, die Schweißgeräte herstellt. Stellen Sie auf Englisch die Produktmerkmale des abgebildeten Gerätes mit den untenstehenden Produktmerkmalen vor.

Produktmerkmale:

a) Leicht tragbar
b) Vierstufige Temperaturauswahl
c) Ideal zum Schweißen dünner Bleche aus Edelstahl oder Aluminium
d) Überhitzungsschutz (automatisches Abschalten)
e) Beinhaltet: Fließdraht, Kontaktierungsklammern, Erdungsklemme
f) Drahtgeschwindigkeitsregelung
g) Funktioniert mit Standardspannung
h) Externe Polaritätsstecker für schnelle Umstellung

Bildquellennachweis

Trackübersicht

An dieser Stelle sind alle Tracks aufgelistet, die sich auf Aufgaben im Workbook beziehen. Die Audio-CD-ROM enthält darüber hinaus alle Tracks des Schülerbuches.

Track	Aufgabe	Laufzeit
A 1.2	Module 1 Refresher course International communication, Aufgabe A 3	00:55
A 1.5	Module 2 The new company, Aufgabe A 1	01:44
A 1.7	Module 2 The new company, Aufgabe A 2	02:02
A 2.12	Module 4 Mechanical engineering – Tools, Aufgabe A 3	02:19
A 1.17	Module 5 Joining and assembly, Aufgabe A 5	01:49
A 2.13	Module 7 Properties of materials, Aufgabe A 7	02:21
A 1.28	Module 8 Electricity basics, Aufgabe A 1	01:43
A 1.32	Module 9 Energy and the environment, Aufgabe A 4	00:43
A 2.2	Module 10 Business trips, Aufgabe A 2	02:41
A 2.5	Module 11 Finding a job in the global village, Aufgabe A 3	03:14
A 2.7	Module 12 Design for manufacturing, Aufgabe A 2	01:17
B 2.10	Module 13 Control technology, Aufgabe A 3	02:02
A 2.11	Module 13 Control technology, Aufgabe A 5	02:32
A 2.14	KMK-Prüfungsvorbereitung 1 Mechanical engineering, Aufgabe 1	01:20
A 2.15	KMK-Prüfungsvorbereitung 2 Welding engineering, Aufgabe 1	02:33
	Gesamtlaufzeit der Audio-CD-ROM	**91:15**